Keys to Exploring Data

Dr Bruce Philip Mull

The

Edition of

Keys to Exploring Data

by

Dr Bruce Philip Mull

Copyright © 2014 by Bruce Philip Mull. All Rights Reserved.
Email: bpmull@hotmail.com

ISBN: 149544855X
ISBN-13: 978-1495448553

Preface

EXPLORATORY DATA ANALYSIS is concerned with describing a sample and determining its characteristics. This book is the first in a series that will cover statistics and probability from an elementary level. Specifically this book covers exploratory data analysis.

One main difference between this book and most is in the treatment of percentiles. The classical methods are based upon assumptions which are often violated in the data. These assumptions are never checked, so the formulas are applied when they should not be. This leads to contradictory calculations when the inverse operations are applied. The method presented in this book shows what those assumptions are and how to develop percentiles when the assumptions are violated. Thus, finding percentiles from measurements and measurements from percentiles give consistent results — the two problems are inverse operations, as they should be. In addition, when the assumptions are met, the method shown reduces to the classical method.

Chapter One gives an introduction to the terminology used in statistics. In addition, it covers sum and product notation — a background topic that does not fit well into any other chapter.

Chapter Two begins with sorting of data. Both qualitative and quantitative data are discussed. It then discusses the grouping of quantitative data: how to determine the number of groups and how to determine the intervals to be used for each group. We introduce the concept of granularity to help determine the number of groups, a concept almost all elementary texts ignore.

Chapter Three covers the basic methods of visualizing data which do not require much processing. It begins with frequency tables as that is the key to the visualizations in the chapter. The visualizations include point diagrams, stem-and-leaf diagrams, bar charts, pie charts, two types of broken line graphs (frequency polygons and ogives), histograms and Pareto diagrams. Both qualitative and quantitative data are covered through visualizations.

Chapter Four begins with a look at the extremes in both the sample and the population. We then look at the effective position and show how this leads to a natural definition of percentile. We look at the two problems of determining the percentile from a measurement and determining the quantile from a percentile. All traditional methods are covered. The problem is that the traditional methods are inconsistent with each other in their solutions to these problems which should be inverses. We also introduce a method which is completely consistent in both directions and show how this method reduces to the traditional methods under specific conditions — conditions never mentioned as being necessary in the traditional approaches. We call this consistent method the proper method for it makes the two problems consistent inverses as they should be. We then cover grouped data and the original definition of percentiles being restricted to the observed limits. We finish with a look at box-and-whiskers diagrams.

Chapter Five looks at the statistics of centrality. It begins with the center, midrange and median before looking at the mode. We then look at the various ways to calculate an average. We do not just give the formulas for these averages, as most books do, but actually show when each should be used. We then give statistics for the most representative value. We look at a wide variety of statistics for both grouped and ungrouped data and how to modify these estimates in the presence of outliers using trimming and Winsorizing.

Chapter Six explores the problem of dispersion and spread. We first quickly cover the range and interquartile range — both of which have been peripherally covered before — and then introduce a variety of statistics to measure the average dispersion, where the spread is determined between observations and as deviations from a fixed point. We then cover the root mean square average and end with the variance and standard deviation and how to deal with outliers in the data. Unlike most elementary texts, we also cover the variation and standard deviance, which differ from the more common statistics in that they use the median instead of the mean as their center of deviation.

Chapter Seven explores skewness and bias in the data. Many statistics of skewness are covered beginning with the tilt and ending with Fisher's skewness coefficient. Some of the statistics covered in this chapter are rarely seen in an elementary text.

Contents

Preface		i
Contents		v
1	**Introduction**	**1**
	1.1 Two Types of Statistics	1
	1.2 Data	2
	1.3 Sampling	4
	1.4 Sum and Product Notation	7
2	**Sorting and Grouping Data**	**19**
	2.1 Sorting Raw Data	19
	2.2 Grouping Data	27
	2.3 Minimum Number of Groups	28
	2.4 Granularity	33
	2.5 Maximum Number of Groups	36
	2.6 Creating the Group Intervals	40
3	**Visualizing Data**	**49**
	3.1 Frequency Tables	49

	3.2	Frequency Table Variations	55
	3.3	Point Diagrams	62
	3.4	Stem-and-Leaf Diagrams	66
	3.5	Bar Charts	72
	3.6	Pie Charts	76
	3.7	Broken Line Graphs	82
	3.8	Histograms and Pareto Diagrams	89

4 Quantiles — 99
	4.1	Extremes	99
	4.2	Effective Position	104
	4.3	Percentile Rank	118
	4.4	Quantiles	130
	4.5	The Sample as Population	145
	4.6	Box-and-Whiskers Diagrams	164

5 Centrality — 181
	5.1	Center of Observations	181
	5.2	Midrange	188
	5.3	Median	191
	5.4	Walsh Sum Median	201
	5.5	Mode	212
	5.6	Types of Averages	218
	5.7	Mean	228
	5.8	Tukey's Trimean	235
	5.9	Trimmed Means	242
	5.10	Winsorized Means	253

6 Data Variability — 271
	6.1	Range & Interquartile Range	271

6.2	Average Absolute Variability	290
6.3	Median Absolute Variability	299
6.4	Average Absolute Deviation	305
6.5	Median Absolute Deviation	315
6.6	Root Mean Square Average	325
6.7	Variation and Standard Deviance	328
6.8	Variance and Standard Deviation	340
6.9	Eliminating Outliers	350
6.10	Standard Z-Scores	362

7 Skewness — 369

7.1	The Tilt	369
7.2	Relative Difference of the Middles	378
7.3	The Method of Differences	384
7.4	Bowley's Skewness Coefficient	392
7.5	Pearson's Skewness Coefficient	399
7.6	Fisher's Skewness Coefficient	405

Chapter 1

Introduction

> *Statistics is concerned with collecting, organizing and analyzing data for description, comparison, inference, prediction or decision making. In short it is the art of learning from data. In this chapter we cover the basic terminology used in statistics.*

1.1 Two Types of Statistics

STATISTICS HAS two uses, hence two branches. Both branches are concerned with collecting, organizing, analyzing data about some group(s) of interest and presenting the information gleaned. **Descriptive statistics** uses the information to discover and describe features of the group. **Inferential statistics** uses it to compare two groups, make inferences or predict things about the group, or make decisions concerning the group.

It starts with a question about some group of interest. The group of interest is the **population**. A **datum** is a fact or quantity that is observed, measured or assumed. The plural of datum is data. Data are gathered to answer this question.

The question is used to create a **study**, a method for obtaining data to solve the question. If the question is about some feature that may or may not be present then a **descriptive** or **exploratory** study is needed and the results will require **descriptive statistics** to answer the question. If the question involves more than one group, or requires that a prediction be made or a conclusion reached, then a **decisive** or **analytical** study is needed and the results will require **inferential statistics** to answer the question.

1.2 Data

When data is gathered directly for the current question it is **primary** data. If it was gathered for some other question but is being reanalyzed for the current question then it is **secondary** data. Both sources have their strengths and weaknesses. This text uses only primary data.

Data itself comes in two main types. **Qualitative** data divides the population into types or categories that cannot be blended together in a meaningful way. **Quantitative** data is numeric and can usually be combined into other numeric quantities which have meaning. Even within each of these data types there are subtle divisions.

Qualitative data is **categorical** or **nominative** when it divides the population into groups by means of labels that have no natural ordering. Qualitative data is **ordinal** when there is a natural ordering. Gender, race or religion are nominal. Alphabets are ordinal. The order in ordinal data may still have an arbitrary aspect. Consider the letters of the Elder Futhark. They can be put in their original order, the order their names would appear in a modern dictionary or the order of the letters they represent.

Quantitative data is **discrete** if it can only take on distinct values. Arbitrary arithmetic operations are not done normally on this type of data as they rarely make sense. They are **interval scale** if differences of the same magnitude are equal, but multiplication and division cannot be interpreted in the normal manner because there is no true zero. Quantitative data is **ratio scale** when all four basic arithmetic operations are interpretable in a normal manner because there is a true zero. Quantitative data is **continuous** when it can accept all real operations, including powers and roots, and there is a smooth flow between values. The data can be real or complex numbers.

Example 1:
There is an apple tree in your yard and you want to determine the average number of leaves per branch. What is your population? What is the nature of the study? What is the nature of the data? What branch of statistics will you need to use?

Answer:
The population is all branches on the apple tree. You wish to describe a feature of these branches — the average number of leaves per branch — so this is a descriptive study. The data are quantitative; specifically discrete data as all data will be nonnegative integers. Because the problem is descriptive, you will use descriptive statistics.

Example 2:
There are five people running for mayor of your city in the upcoming election and you wish to determine which of them is most likely to

win. What is your population? What is the nature of the study? What is the nature of the data? What branch of statistics will you need to use?

Answer:

Your population is all voters in your city. You wish to predict the winner in the election so this is an analytical study. The data are qualitative; specifically, nominal data as each datum will be the name of the person the voter will vote for in the election. Because the problem is to make a prediction, you will use inferential statistics.

Exercises 1.2

1. Your favorite cereal is *Oats 'n' Raisins*. On the front of the box it claims there are at least 200 raisins per box and you wish to determine whether this is true. What is your population? What is the nature of the study? What is the nature of the data? What branch of statistics will you need to use?

2. You are curious to find out what the eight most commonly used letters are in the fiction of H. P. Lovecraft that are not collaborations. What is your population? What is the nature of the study? What is the nature of the data? What branch of statistics will you need to use?

1.3 Sampling

WE WILL NOT GO into great detail as this is an introductory book. A **sample** is a nonempty subset of individuals from the population.

1.3. SAMPLING

If the sample includes all members of the population then it is a **census**. Basically, **sampling** deals with the way a sample is chosen.

An **observation** is made about each individual in the sample. These observations are recorded as the **data** from the study. The observations may be quantitative or qualitative, depending upon what exactly is desired from the study.

How you create a sample can be as important as the recorded observations themselves. A sample is **random** if every individual in the population has the same chance to be selected into the sample as every other individual. Samples which are not random are **biased**. However, bias in the sample may or may not lead to bias in the result, but *is* more likely to do so.

Another technique is **every n^{th} one**. In this technique you choose a number n, not necessarily randomly, which gives the frequency of sampling. Often, you also randomly choose another number which is the first individual that will be chosen. If you do not choose a starting number then n is the first value chosen (n, $2n$, $3n$, ...); otherwise, if you start at k then the chosen values are k, $k+n$, $k+2n$, $k+3n$, etc.

If your population is divided into categories and you need to make certain that every category is represented then you may want to do **stratified** sampling. Here you ensure that at least one member from each category is chosen, but do random or every n^{th} one sampling within each category. It is not necessary for the same n to be used in each category.

The main thing you want from a sample is for it to be representative. A sample is **representative** if the results from the sample agree with results that would be obtained from a census. There is no way to *guarantee* that a sample which is not a census is representative. If a sample is random then there is no reason to believe that the results will reflect any particular minority. For this reason, random samples are *assumed* to be representative. If you do not order the individuals in the population before picking them then every n^{th} one also tends to produce representative samples.

If some categories are rare then you may wish to oversample those categories. **Oversampling** is a biased technique which makes sure that too many individuals are chosen from some, possibly rare, category. When the weights are chosen so that each category is brought back into line with its proportion in the population then it is *assumed* that the bias has been removed.

Example 1:
You pick the two lowest branches and count the leaves on each of them to determine the average number of leaves per branch. Is your sample random or biased? Is the result of your study representative?

Answer:
   Not every branch has the same chance to be in your sample, so this is a biased technique. Lower branches are normally larger, and hence probably have more leaves, than upper branches. Therefore, the sample is probably not representative.

Example 2:
You obtain the list of registered voters from City Hall and using a random number table choose a value n, then contact every n^{th} voter until you receive 100 valid responses. Is your sample random or biased? Is the result of your study representative?

Answer:
   You <u>tried</u> to make it random, but it isn't. The voting list is almost assuredly alphabetized; thus, people appearing at the back of the list are

not as likely to be chosen as those at the front of the list. Because the sampling technique is biased, it is impossible to guess whether the results obtained will be representative. Therefore, we assume it will not be.

Exercises 1.3

1. You go to the manufacturer of *Oats 'n' Raisins* cereal. The person in charge takes you on a tour of the facility and you impulsively select ten boxes of the cereal, one at a time, from a randomly chosen production line. You count the number of raisins in each box. Is your sample random or biased? Is the result of your study representative?
2. You pick your favorite Lovecraft story, *At the Mountains of Madness*, randomly choose a page, and tally how many times each letter appears. Is your sample random or biased? Is the result of your study representative?

1.4 Sum and Product Notation

MANY FORMULAS in statistics and other disciplines require sums and/or products. In this section we look at the notation for sums and products.

Summation Notation

SUPPOSE YOU wished to sum up the numbers a_1, a_2, a_3, ..., a_{99}, a_{100}. You *could* write this as

$$a_1 + a_2 + a_3 + \cdots + a_{99} + a_{100}.$$

The problems are 1) you must include enough terms to make sure that everyone knows what pattern is meant by the ellipsis (...); 2) it would be difficult to use this as part of a larger expression as it could quickly use up the line, and more; 3) this type of thing is quite common.

To address these problems a notation for sums was developed. The **summation notation** or **sigma notation** uses a large majuscule Greek letter S (Σ), which stands for the word sum. The sum is controlled by an **index variable** or **control variable**. Traditionally, the index variable is one of the letters i, through n, except ℓ is used instead of l. This is not a requirement, however, and you are free to use any symbol you desire.

The control variable can be used as an index, so we could write the above as

$$\sum_{i=1}^{100} a_i.$$

However, if we wanted to sum the odd integers up to 99, we could write

$$\sum_{j=1}^{50} (2j-1)$$

and make the control variable part of the expression of the sum.

The following pseudocode shows how the notation works. We let sum represent the sum in the pseudocode.

```
sum = 0
for i = start to end by 1
  sum = sum + expr(i)
return sum
```

1.4. SUM AND PRODUCT NOTATION

The sum is initialized to zero. The control value is assigned its start value. If the current value of the control variable is less than or equal to the end value then the expression is evaluated for the current iteration and the sum is augmented by that amount. After this is done, the control variable is incremented by one and tested against the end value. Once the value of the control variable exceeds the end value, the loop is exited and the final sum is returned.

Example 1:
Evaluate the following sums:

1. $\sum_{i=1}^{5} i^2$

2. $\sum_{j=3}^{6} 2^j$

3. $\sum_{k=3}^{7} k \cdot (k-1)$

Answer:

1. $\sum_{i=1}^{5} i^2 = 1^2 + 2^2 + 3^2 + 4^2 + 5^2 = 1 + 4 + 9 + 16 + 25 = 55.$

2. $\sum_{j=3}^{6} 2^j = 2^3 + 2^4 + 2^5 + 2^6 = 8 + 16 + 32 + 64 = 120.$

3. $\sum_{k=3}^{7} k \cdot (k-1) = 3 \cdot 2 + 4 \cdot 3 + 5 \cdot 4 + 6 \cdot 5 + 7 \cdot 6$
$= 6 + 12 + 20 + 30 + 42 = 110.$

The control variable does not need to appear in the expression at all. And if the start value is greater than the end value, we have an **empty sum**. The pseudocode suggests that an empty sum should equal zero and by definition this is true.

Example 2:
Evaluate each of the following sums:

1. $\sum_{\ell=1}^{4} 6$

2. $\sum_{m=5}^{2} 99999^{999m}$

Answer:

1. $\sum_{\ell=1}^{4} 6 = \underbrace{6}_{\ell=1} + \underbrace{6}_{\ell=2} + \underbrace{6}_{\ell=3} + \underbrace{6}_{\ell=4} = 24.$

2. $\sum_{m=5}^{2} 99999^{999m} = 0$ (as $5 > 2$)

Theorem 1.1. *Suppose c is any real constant value and m and n are integers with $m \leq n$. Then*

$$\sum_{i=m}^{n} c = c \cdot (n - m + 1).$$

Proof. Think of the index values as being $m+0, m+1, \ldots, m+(n-m)$. There are $n - m + 1$ terms, each with value c. □

1.4. SUM AND PRODUCT NOTATION

Theorem 1.2. *Suppose a_m, a_{m+1}, ..., a_n are real numbers and c is any real constant. Then*

$$\sum_{i=m}^{n} c \cdot a_i = c \cdot \sum_{i=m}^{n} a_i.$$

Proof. We expand

$$\sum_{i=m}^{n} c \cdot a_i = c \cdot a_m + c \cdot a_{m+1} + \cdots + c \cdot a_n$$
$$= c \cdot (a_m + a_{m+1} + \cdots + a_n)$$
$$= c \cdot \sum_{i=m}^{n} a_i. \qquad \square$$

Theorem 1.3. *Suppose a_1, a_2, ..., a_{m-1}, a_m, a_{m+1}, ..., a_n are real numbers. Then*

$$\sum_{i=1}^{m-1} a_i + \sum_{i=m}^{n} a_i = \sum_{i=1}^{n} a_i.$$

Proof. The left side is

$$(a_1 + a_2 + \cdots + a_{m-1}) + (a_m + a_{m+1} + \cdots + a_n).$$

Being finite, the associative property says can be written as

$$a_1 + a_2 + \cdots + a_{m-1} + a_m + a_{m+1} + \cdots + a_n.$$

Which is exactly what one gets from the right side sum. $\square$

Theorem 1.3 assumes $m \leq n$. In fact this is necessary as it does *not* hold for $m > n$. The same condition holds in the following corollary.

Corollary 1.3.1. *Suppose $a_1, a_2, \ldots, a_{m-1}, a_m, a_{m+1}, \ldots, a_n$ are real numbers. Then*

$$\sum_{i=m}^{n} a_i = \sum_{i=1}^{n} a_i - \sum_{i=1}^{m-1} a_i.$$

Proof. All sums are finite, so the commutative, associative, additive inverse, and zero sum properties apply. □

Theorem 1.4. *Suppose $a_1, a_2, \ldots, a_n$ and $b_1, b_2, \ldots, b_n$ are real numbers. Then*

$$\sum_{i=1}^{n}(a_i + b_i) = \sum_{i=1}^{n} a_i + \sum_{i=1}^{n} b_i$$

and

$$\sum_{i=1}^{n}(a_i - b_i) = \sum_{i=1}^{n} a_i - \sum_{i=1}^{n} b_i$$

Proof. Apply the commutative and associative properties together with Theorem 1.2 and the definition of subtraction. □

Product Notation

JUST AS SUMS OFTEN show up in expressions, so do products. Suppose you wished to multiply the numbers $a_1, a_2, \ldots, a_{100}$. You could write this as

$$a_1 \cdot a_2 \cdots \cdots a_{100},$$

which has the same pitfalls as writing sums. Because of this, **product notation** or **pi notation** was developed. The **pi notation** uses a large majuscule Greek letter P (Π), which stands for the word product. The notation is similar to that for sums. The product is

1.4. SUM AND PRODUCT NOTATION

controlled by an index variable in the same way. The same traditional letters are used for the control variable of a product as are used for sums, which again is not a requirement, so you are free to use any symbol you like.

The control variable can be used as an index, so the above would be written

$$\prod_{j=1}^{100} a_j.$$

However, if we wanted the product of the even numbers from 2 to 100 we could write this as

$$\prod_{k=1}^{50} 2k$$

and make the control variable part of the expression of the product.

The following pseudocode shows how the notation works. We let **prd** represent the product in the pseudocode.

```
prd = 1
for i = start to end by 1
  prd = prd * expr(i)
return prd
```

The product is initialized to one. The control value is assigned its start value. If the current value of the control variable is less than or equal to the end value then the expression is evaluated for the current iteration and the product is scaled by that amount. After this is done, the control variable is incremented by one and tested against the end value. Once the value of the control variable exceeds the end value, the loop is exited and the final product is returned.

Example 3:
Evaluate each of the following products.

1. $\prod_{i=1}^{5} i^2$

2. $\prod_{j=3}^{6} 2^j$

3. $\prod_{k=3}^{7} k \cdot (k-1)$

Answer:

1. $\prod_{i=1}^{5} i^2 = 1^2 \cdot 2^2 \cdot 3^2 \cdot 4^2 \cdot 5^2 = 1 \cdot 4 \cdot 9 \cdot 16 \cdot 25 = 14400.$

2. $\prod_{j=3}^{6} 2^j = 2^3 \cdot 2^4 \cdot 2^5 \cdot 2^6 = 2^{3+4+5+6} = 2^{18} = 262144.$

3. $\prod_{k=3}^{7} k \cdot (k-1) = (3 \cdot 2) \cdot (4 \cdot 3) \cdot (5 \cdot 4) \cdot (6 \cdot 5) \cdot (7 \cdot 6)$
$= 6 \cdot 12 \cdot 20 \cdot 30 \cdot 42 = 1814400.$

The control variable does not need to appear in the expression at all. And if the start value is greater than the end value, we have an **empty product**. The pseudocode suggests that an empty product should equal one and by definition this is true.

Example 4:
Evaluate each of the following products:

1.4. SUM AND PRODUCT NOTATION

1. $\displaystyle\prod_{\ell=1}^{4} 6$

2. $\displaystyle\prod_{m=5}^{2} 99999^{999m}$

Answer:

1. $\displaystyle\prod_{\ell=1}^{4} 6 = \underbrace{6}_{\ell=1} \cdot \underbrace{6}_{\ell=2} \cdot \underbrace{6}_{\ell=3} \cdot \underbrace{6}_{\ell=4} = 6^4 = 1296.$

2. $\displaystyle\prod_{m=5}^{2} 99999^{999m} = 1$ (as $5 > 2$)

Theorem 1.5. *Suppose c is any real constant value and m and n are integers with $m \leq n$. Then*

$$\prod_{i=m}^{n} c = c^{n-m+1}.$$

Proof. Think of the index values as being $m+0$, $m+1$, ..., $m+(n-m)$. There are $n-m+1$ factors, each with value c. □

Theorem 1.6. *Suppose a_m, a_{m+1}, ..., a_n are real numbers and c is any real constant. Then*

$$\prod_{i=m}^{n} c \cdot a_i = c^{n-m+1} \cdot \prod_{i=m}^{n} a_i.$$

Proof. We expand

$$\prod_{i=m}^{n} c \cdot a_i = (c \cdot a_m) \cdot (c \cdot a_{m+1}) \cdot \ldots \cdot (c \cdot a_n)$$
$$= c^{n-m+1} \cdot (a_m \cdot a_{m+1} \cdot \ldots \cdot a_n)$$
$$= c^{n-m+1} \cdot \prod_{i=m}^{n} a_i. \qquad \square$$

Theorem 1.7. *Suppose a_1, a_2, ..., a_{m-1}, a_m, a_{m+1}, ..., a_n are real numbers. Then*

$$\prod_{i=1}^{m-1} a_i \cdot \prod_{i=m}^{n} a_i = \prod_{i=1}^{n} a_i.$$

Proof. The left side is

$$(a_1 \cdot a_2 \cdot \ldots \cdot a_{m-1}) \cdot (a_m \cdot a_{m+1} \cdot \ldots \cdot a_n).$$

Being finite, the associative property says can be written as

$$a_1 \cdot a_2 \cdot \ldots \cdot a_{m-1} \cdot a_m \cdot a_{m+1} \cdot \ldots \cdot a_n.$$

Which is exactly what one gets from the right side sum. $\qquad \square$

Theorem 1.7 assumes $m \leq n$. In fact this is required because the theorem does not hold in general when $m > n$. The same assumptions and conditions apply to the following corollary.

Corollary 1.7.1. *Suppose a_1, a_2, ..., a_{m-1}, a_m, a_{m+1}, ..., a_n are real numbers. Then*

$$\prod_{i=m}^{n} a_i = \frac{\prod_{i=1}^{n} a_i}{\prod_{i=1}^{m-1} a_i}.$$

1.4. SUM AND PRODUCT NOTATION

Proof. All products are finite, so the commutative, associative, multiplicative inverse, and unit product properties apply. □

Theorem 1.8. *Suppose $a_1, a_2, \ldots, a_n$ and $b_1, b_2, \ldots, b_n$ are real numbers. Then*

$$\prod_{i=1}^{n}(a_i \cdot b_i) = \prod_{i=1}^{n} a_i \cdot \prod_{i=1}^{n} b_i$$

and

$$\prod_{i=1}^{n} \frac{a_i}{b_i} = \frac{\prod_{i=1}^{n} a_i}{\prod_{i=1}^{n} b_i}$$

Proof. Apply the commutative and associative properties together with the definition of division. □

Exercises 1.4

1. Given $\sum_{i=1}^{n} i = \frac{n \cdot (n+1)}{2}$ and $\sum_{i=1}^{n} 1 = n$, show that

$$\sum_{i=1}^{n}(2i - 1) = n^2.$$

2. Evaluate $\sum_{i=17}^{30} i$. [Hint: Use Corollary 1.3.1 and the given information of Exercise 1, above.]

3. Given $\prod_{j=1}^{n} j = n!$. Prove $\prod_{j=1}^{n} j^2 = (n!)^2$. [Hint: Use Theorem 1.8.]

4. Evaluate $\prod_{j=6}^{12} j^2$. [Hint: Use Corollary 1.7.1 and the result of Exercise 3, above.]

Chapter 2

Sorting and Grouping Data

In this chapter we begin our discussion of descriptive statistics by looking at sorting and grouping of data. Both qualitative and quantitative data are discussed. However, only quantitative data is considered for the general grouping techniques.

2.1 Sorting Raw Data

STATISTICS ARE numerical operations on data. **Raw data** is data in the order it was gathered, when absolutely no processing has been done with it. The calculation of a statistic usually requires that the data at least be sorted. Often, even statistics that do not require sorting are easier to calculate when sorting has been done. Therefore, we begin with sorting.

Suppose a sample has n observations. We let $x_1, x_2, \ldots, x_n$ represent the raw data. When the data are sorted this generates $x_{(1)}$, $x_{(2)}, \ldots, x_{(n)}$ — the sorted data. We use parentheses to indicate the sorted order. Thus, $x_{(1)}$ is the value of the first ordered datum and $x_{(n)}$ is the value of the last sorted datum. In general, $x_{(i)}$ is the value of the datum which ended in the ith position in the sorted list of data.

It may happen that some values are in the same position in both the sorted and unsorted ordering. Depending upon the number of possible values in comparison to the number of observations this may or may not be rare.

Numeric Data

QUANTITATIVE DATA is normally sorted from smallest to largest. There are a number of sorting routines that one can use. When sorting by hand, I tend to sort each row using a double selection sort and then use a merge sort to get the final results.

Example 1:
The following set of values are the weights, in grams of 50 randomly selected oysters shucked in 2013.

9.20	9.98	9.53	9.47	6.67	7.37	11.59	8.12	7.40	6.97
11.09	10.81	10.24	6.63	10.63	10.86	13.18	9.39	8.42	8.61
9.21	5.76	10.81	7.97	7.71	11.97	6.91	9.55	11.17	6.91
6.37	9.01	7.03	6.42	11.02	7.04	10.63	6.10	10.62	6.53
6.36	8.53	11.31	8.02	12.07	10.50	8.04	8.62	8.14	6.74

Sort the data from smallest to largest:

Answer:
Scanning across the first row, I see 6.67 is smallest and 11.59 is largest:

2.1. SORTING RAW DATA

    6.67   9.20   9.98   9.53   9.47   7.37   8.12   7.40   6.97   11.59

Now, 6.97 is smallest and 9.98 is largest:

    6.67   6.97   9.20   9.53   9.47   7.37   8.12   7.40   9.98   11.59

Now, the smallest and largest are 7.37 and 9.53, respectively:

    6.67   6.97   7.37   9.20   9.47   8.12   7.40   9.53   9.98   11.59

This makes 7.40 smallest and 9.47 largest:

    6.67   6.97   7.37   7.40   9.20   8.12   9.47   9.53   9.98   11.59

Swapping the middle pair sorts the first row.

    6.67   6.97   7.37   7.40   8.12   9.20   9.47   9.53   9.98   11.59

Doing this for each of the rows gives:

6.67	6.97	7.37	7.40	8.12	9.20	9.47	9.53	9.98	11.59
6.63	8.42	8.61	9.39	10.24	10.63	10.81	10.86	11.09	13.18
5.76	6.91	6.91	7.71	7.97	9.21	9.55	10.81	11.17	11.97
6.10	6.37	6.42	6.53	7.03	7.04	9.01	10.62	10.63	11.02
6.36	6.74	8.02	8.04	8.14	8.53	8.62	10.50	11.31	12.07

To finish the sort, I look at the first remaining entry of each row and pick the smallest, crossing it off, continuing until I have the final first row, then the final second row, and so on. Below we show the first row done. The next one that will be picked is the second 6.91, which will start the second row.

    5.76   6.10   6.36   6.37   6.42   6.53   6.63   6.67   6.74   6.91

~~6.67~~	6.97	7.37	7.40	8.12	9.20	9.47	9.53	9.98	11.59
~~6.63~~	8.42	8.61	9.39	10.24	10.63	10.81	10.86	11.09	13.18
~~5.76~~	~~6.91~~	6.91	7.71	7.97	9.21	9.55	10.81	11.17	11.97
~~6.10~~	~~6.37~~	~~6.42~~	~~6.53~~	7.03	7.04	9.01	10.62	10.63	11.02
~~6.36~~	~~6.74~~	8.02	8.04	8.14	8.53	8.62	10.50	11.31	12.07

Continuing on with each row produces the final sorted table, presented below.

5.76	6.10	6.36	6.37	6.42	6.53	6.63	6.67	6.74	6.91
6.91	6.97	7.03	7.04	7.37	7.40	7.71	7.97	8.02	8.04
8.12	8.14	8.42	8.53	8.61	8.62	9.01	9.20	9.21	9.39
9.47	9.53	9.55	9.98	10.24	10.50	10.62	10.63	10.63	10.81
10.81	10.86	11.02	11.09	11.17	11.31	11.59	11.97	12.07	13.18

If you examine the original and sorted order you will find that not a single datum is in the same position in both tables. That is, for all $i = 1, \ldots, 50$, $x_{(i)} \neq x_i$.

Nonnumeric Data

WHEN THE DATA are nonnumeric then there may or may not be a natural ordering for the data. There are three basic ways to sort nonnumeric data:

1. use the natural order,

2. sort alphabetically,

3. sort in encountered order.

We will look at examples for all three sorting orders.

Example 2:

The original English alphabet, the alphabet of the angli saxonii (Angelic Saxons — blonde Teutons), is the futhorc (more properly fuþorc) which was derived from the elder futhark (fuþark). In both cases the name is derived from the first six letters. The order of this alphabet is

ᚠᚢᚦᚩᚱᚳᚷᚹᚻᚾᛁᛄᛇᛈᛉᛋᛏᛒᛖᛗᛚᛝᛟᛞᚪᚫᚣᛡᛠᚸ

Some groups added ᛞ, for the allophone of ᚦ, and ᛡ for the modern ck combination.

2.1. SORTING RAW DATA

A bag containing the 33 runes of the fuþorc etched onto wooden tiles is shaken and a tile is withdrawn, the rune is recorded and the tile is replaced and the bag remixed. The procedure is repeated 105 times to produce the following raw data:

[runic data table]

Sort these data into their original order as listed above.

Answer:

Unless you are familiar with this alphabet, it could take a while as some letters look similar to others — the o (ᚯ), a (ᚪ) and æ (ᚫ), or the hard g (ᚷ) and soft g (ᚸ). Here is the result of the sort.

[sorted runic data table]

Comparing the sorted and original orders, we see that $x_{(38)} = x_{38} = $ ᛁ (i) and $x_{(102)} = x_{102} = $ ᛥ (st). No other letters are in the same position.

Example 3:

A bag of 500 balloons in assorted colors was dumped into a box and the box was shaken. Balloons were drawn out one at a time and the color of the balloon was recorded. The balloons were not replaced into the box, but between draws the box was shaken. The result of 25 withdrawals is given in the table below.

orange	blue	blue	red	yellow
yellow	green	green	red	yellow
blue	yellow	orange	orange	red
blue	orange	green	blue	red
orange	red	blue	green	yellow

Sort the data in alphabetic order.

Answer:

There are five colors: blue, green, orange, red and yellow. Sorting the names in this order gives

blue	blue	blue	blue	blue
blue	green	green	green	green
orange	orange	orange	orange	orange
red	red	red	red	red
yellow	yellow	yellow	yellow	yellow

Comparing the sorted and unsorted data $x_{(2)} = x_2 =$ blue, $x_{(3)} = x_3 =$ blue, $x_{(7)} = x_7 =$ green, $x_{(8)} = x_8 =$ green, $x_{(13)} = x_{13} =$ orange, $x_{(14)} = x_{14} =$ orange, $x_{(20)} = x_{20} =$ red and $x_{(25)} = x_{25} =$ yellow. Eight of the data are in the same position as in the original ordering. Note: The manufacturer claims on the bag that, on average, every color appears equally.

Example 4:

A large bag of candy-coated chocolate drops has 1000 pieces in it. After shaking the bag vigorously to make sure that the colors are mixed, you scoop out 48 pieces of candy and record the color of the candy coating. The order in which the pieces were counted and tabulated is shown below.

2.1. SORTING RAW DATA

orange	brown	yellow	orange	brown	red	yellow	green
blue	red	brown	blue	blue	brown	blue	red
brown	blue	brown	yellow	yellow	brown	blue	orange
orange	yellow	brown	red	yellow	orange	orange	orange
green	brown	red	brown	blue	blue	brown	red
yellow	blue	green	brown	brown	red	orange	red

Sort the data in encountered order: orange, brown, yellow, red, green and blue.

Answer:
  The sorted data is

orange	orange	orange	orange	orange	orange	orange	orange
brown	brown	brown	brown	brown	brown	brown	brown
brown	brown	brown	brown	brown	yellow	yellow	yellow
yellow	yellow	yellow	yellow	red	red	red	red
red	red	red	red	green	green	green	blue
blue	blue	blue	blue	blue	blue	blue	blue

Comparing the sorted and unsorted data $x_{(1)} = x_1 =$ orange, $x_{(4)} = x_4 =$ orange, $x_{(11)} = x_{11} =$ brown, $x_{(14)} = x_{14} =$ brown, $x_{(17)} = x_{(17)} =$ brown, $x_{(19)} = x_{19} =$ brown, $x_{(26)} = x_{26} =$ yellow, $x_{(35)} = x_{35} =$ red and $x_{(42)} = x_{42} =$ blue. Nine of the data are in the same position as in the original ordering. There is no claim by the manufacturer that all colors are equally likely.

When data are sorted in encountered order, one must have $x_{(1)} = x_1$. After all, the first datum must be in some category and this *must* be the first time that category was encountered. This is the only ordering that guarantees that some datum is in the same position in both the sorted and unsorted data.

Exercises 2.1

1. One day while walking through a field you notice there seem to be a lot of stones lying on the ground. Curious about what is the average size of a stone in the field you randomly collect 30 stones. You weigh each stone and record its weight in grams. The following table shows the raw data.

110	126	75	110	126	94
90	112	111	139	131	92
102	103	96	75	67	101
104	111	100	65	92	93
106	104	93	115	109	95

   Sort these data from smallest to largest.

2. The scores, in the order graded, on a recent midterm examination in Statistics, a class with 30 students, are given below.

70	80	99	98	85	89
87	79	83	38	69	70
60	69	78	40	75	56
70	51	99	69	95	86
57	53	47	50	55	81

   Sort these data from smallest to largest.

3. A bag of candy-coated chocolate drops has 42 pieces of candy in it. You recorded the colors of each piece as it was removed from the bag and the result is given in the follow table.

red	brown	yellow	brown	green	yellow	yellow
brown	brown	brown	brown	red	brown	red
brown	brown	blue	brown	yellow	brown	blue
green	brown	yellow	yellow	brown	brown	brown
yellow	orange	orange	yellow	yellow	red	brown
brown	blue	red	blue	yellow	yellow	red

   Sort these data alphabetically by the name of the color.

4. A pair of dice, one red and one green, is rolled in the following manner. On the first toss both dice are tossed. Then the red die is tossed while the green die keeps its value to get the second sum. Then the green die is tossed while the red die keeps its value. You continue tossing alternately the red then the green, die and generate the following table.

8	11	8	6	10	9	9
7	2	5	10	11	8	5
3	6	9	7	7	9	10
7	3	3	7	9	8	7
5	4	8	11	11	10	10

Sort these data in encountered order. [Bonus: In the original toss the red die was 3 and the green die was 5. Try to determine the sequence of red, green, alternating tosses.]

5. You toss a coin 20 times and record whether the obverse side (heads) or reverse side (tails) fell up. The result of these twenty tosses is shown below. In the table, H means heads and T means tails.

T	H	T	T	H
H	T	T	H	T
H	H	T	T	H
T	H	H	T	H

Assume T is less than H and sort accordingly.

2.2 Grouping Data

QUALITATIVE is already grouped so unless the data can be classified into broader categories there is very little that can be done.

Letters can be broadly categorized as consonants, vowels and semivowels. If that is too few groups, then consonants can be classified as allophones, fricatives, multiples, mutes, nasals, sibilants and spirants.

Ordinal data can be converted into discrete data and then if there are enough distinct values, could be grouped using the discrete values. But this must be done carefully otherwise the mapping from ordinal to discrete could hide more than it reveals. For example, the difference between "strongly agree" and "agree" may not be the same as the difference between "disagree" and "strongly disagree." The point is that grouping nonnumeric data should be done with caution.

Numeric data have the natural order induced by the real numbers. Normally, placing them into intervals still makes the results meaningful. If it doesn't then perhaps the data are actually categorical and only pretending to be numeric. One might be better off replacing such "numbers" with "names" to make them truly nominal.

A word of caution: numeric data *always* lose precision through grouping as you no longer have access to the original values. Thus knowing that some observation x was in the interval $[0, 10]$ does not tell you whether x was close to 0, close to 10, or close to 5. When used in calculations it will be assumed close to 5 and this will have an impact on the precision of your statistical estimate. If the grouping is done correctly then this loss in precision will be minimal and balanced by ease of computation.

2.3 Minimum Number of Groups

From this point on we will assume that the data are numeric and that grouping the data into intervals does not destroy the meaning-

2.3. MINIMUM NUMBER OF GROUPS

fulness of the data. There are two goals to grouping data: 1) make the data easier to use to calculate specific statistics and 2) enable the data to be shown visually in a manner that allows its features to be seen. There is a third criterion that is harder to classify: Make the visualization of the data look nice. We begin by considering the minimum number of groups that are needed.

We assume that the sample size is n; that is, there are n observations in the sample. We let g be the number of groups. The minimum number of groups is based only on the size of the sample. There are three commonly used bounds for g:

1. $g \geq \lceil \lg(n) \rceil$
2. $g \geq \lceil \ln(n+1) \rceil$
3. $g \geq 2 + \lceil \log(n) \rceil$.

The notation $\lg x$ refers to logarithms taken to the base 2. The notation $\ln x$ refers to logarithms taken to the base e, where the value of e is e $\approx$ 2.718281828459. The notation $\log x$ refers to logarithms taken to the base 10. Most scientific calculators have keys for $\ln x$ and $\log x$. To get logarithms to base 2 you can use the relation

$$\lg x = \frac{\ln x}{\ln 2} = \frac{\log x}{\log 2}.$$

Some business calculators have either $\ln x$ or $\log x$, but not both. The following relations can be used to get the other

$$\log x = \frac{\ln x}{\ln 10} \quad \text{and} \quad \ln x = \frac{\log x}{\log(2.718281828459)}$$

The notation $\lceil x \rceil$ is the **least integer function** and means to take the smallest integer that is greater than or equal to x.

For reasonably sized samples these values are listed from largest to smallest. Personally, if only one is used, I recommend the middle one. This is the one illustrated in the examples.

Some people place additional restrictions on the groups.

- Many people say that $g \geq 3$ or even $g \geq 5$ should always hold and would reject any minimum less than these.

- For most samples (but not all) an odd number of groups looks nicer than an even number of groups. Because of this some people insist that the number of intervals be odd.

Your instructor may insist on one or both of these and may have additional restrictions he/she insists be followed.[1]

Example 1:
Determine the minimum number of groups needed for the oyster data of Example 1, §2.1, which starts on page 20.

Answer:
There are 50 observations in the sample. By the middle formula

$$g \geq \lceil \ln(51) \rceil \approx \lceil 3.9318 \rceil = 4.$$

We conclude that at least 4 groups should be used.

For comparison,

$$\lceil \lg(50) \rceil \approx \lceil 5.6439 \rceil = 6 \text{ and } 2 + \lceil \log(50) \rceil \approx 2 + \lceil 1.6990 \rceil = 2 + 2 = 4.$$

---

[1] My advice is follow the rules set by your instructor as he/she may have valid reasons for imposing the additional restrictions.

2.3. MINIMUM NUMBER OF GROUPS

Example 2:
A certain event occurs about one-third of the time. The table below gives the number of tries needed to get two occurrences of this event. The experiment was repeated 100 times with the following results.

2	11	5	3	5	7	5	6	3	28
4	16	10	14	5	7	12	10	8	3
3	4	3	5	5	2	5	17	2	14
6	2	9	9	4	6	6	6	11	4
2	5	3	8	2	5	4	6	7	10
9	15	12	10	8	14	4	5	5	4
3	11	4	9	5	4	4	5	3	7
14	4	14	8	2	6	4	5	4	6
2	3	3	3	5	9	4	10	3	11
3	3	6	4	3	2	3	5	10	5

Determine the minimum number of groups needed for this data.

Answer:
By the middle formula, the minimum number of groups should be

$$g \geq \lceil \ln(101) \rceil = \lceil 4.6151 \rceil = 5.$$

We conclude there should be at least 5 groups.

Exercises 2.3

1. One day while walking through a field you notice there seem to be a lot of stones lying on the ground. Curious about what the average size of a stone is in the field you randomly collect 30 stones. You weigh each stone and record its weight in grams. The following table shows the raw data.

110	126	75	110	126	94
90	112	111	139	131	92
102	103	96	75	67	101
104	111	100	65	92	93
106	104	93	115	109	95

Determine the minimum number of groups needed for this data.

2. The scores, in the order graded, on a recent midterm examination in Statistics, a class with 30 students, are given below.

70	80	99	98	85	89
87	79	83	38	69	70
60	69	78	40	75	56
70	51	99	69	95	86
57	53	47	50	55	81

Determine the minimum number of groups needed for this data.

3. A pair of dice, one red and one green, is rolled in the following manner. On the first toss both dice are tossed. Then the red die is tossed while the green die keeps its value to get the second sum. Then the green die is tossed while the red die keeps its value. You continue tossing alternately the red then the green, die and generate the following table.

8	11	8	6	10	9	9
7	2	5	10	11	8	5
3	6	9	7	7	9	10
7	3	3	7	9	8	7
5	4	8	11	11	10	10

Determine the minimum number of groups needed for this data.

2.4 Granularity

IF YOU USE TOO few groups then you may miss important features. On the other hand, if you use too many groups then you may see things which are not really there. It is therefore just as important for you to determine the maximum number of groups as it is the minimum number. Here it is not only sample size that is important, but also the granularity of the sample. The idea is that every interval should have at least two distinct values in it, but that the intervals not be so small as to cause gaps between intervals — empty intervals. The **granularity**, denoted G, is the minimum spacing that will not cause gaps between the group intervals. It is important that the data be sorted to be able to determine the granularity.

1. Sort the data and determine the list of distinct values.

2. Determine the minimum difference between consecutive distinct values and call this d.

3. Determine the maximum difference between consecutive distinct values and call this D.

4. Set $G = \max(2 \cdot d, D)$.

Example 1:
Determine the granularity of the oyster data of Example 1, §2.1, which started on page 20.

Answer:
From the sorted data generated in that example, the distinct values are

5.76	6.10	6.36	6.37	6.42	6.53	6.63	6.67	6.74	6.91
6.97	7.03	7.04	7.37	7.40	7.71	7.97	8.02	8.04	8.12
8.14	8.42	8.53	8.61	8.62	9.01	9.20	9.21	9.39	9.47
9.53	9.55	9.98	10.24	10.50	10.62	10.63	10.81	10.86	11.02
11.09	11.17	11.31	11.59	11.97	12.07	13.18			

The minimum difference is $d = 0.01$ which first occurs as $6.37 - 6.36$. The maximum difference is $D = 0.43$ which occurs as $9.98 - 9.55$. The granularity is
$$G = \max(2(0.01), 0.43) = 0.43.$$

Example 2:
Determine the granularity of the number of tries to get two occurrences of an event from Example 2, §2.3, which started on page 30.

Answer:
We begin by sorting the data. this gives

2	2	2	2	2	2	2	2	2	3
3	3	3	3	3	3	3	3	3	3
3	3	3	3	3	4	4	4	4	4
4	4	4	4	4	4	4	4	4	4
5	5	5	5	5	5	5	5	5	5
5	5	5	5	5	5	5	6	6	6
6	6	6	6	6	6	7	7	7	7
8	8	8	8	9	9	9	9	9	10
10	10	10	10	10	11	11	11	11	12
12	14	14	14	14	14	15	16	17	28

From this we determine the distinct values are

| 2 | 3 | 4 | 5 | 6 | 7 | 8 | 9 |
| 10 | 11 | 12 | 14 | 15 | 16 | 17 | 28 |

The minimum difference is $d = 1$ which first occurs as $3 - 2$. The maximum difference is $D = 11$ which occurs as $28 - 17$.
$$G = \max(2(1), 11) = 11.$$

2.4. GRANULARITY

Note: This maximum difference is almost half of the distance from 2 to 28. In this case, the granularity might best be compared to the second highest difference, $D = 2$ which occurs at $14 - 12$ which would give a more reasonable granularity of 2, but then we are eliminating 28 from this calculation and we would regard 17 as being the top for this granularity.

Exercises 2.4

1. One day while walking through a field you notice there seem to be a lot of stones lying on the ground. Curious about what the average size of a stone is in the field you randomly collect 30 stones. You weigh each stone and record its weight in grams. The following table shows the raw data.

110	126	75	110	126	94
90	112	111	139	131	92
102	103	96	75	67	101
104	111	100	65	92	93
106	104	93	115	109	95

   Determine the granularity of this data.

2. The scores, in the order graded, on a recent midterm examination in Statistics, a class with 30 students, are given below.

70	80	99	98	85	89
87	79	83	38	69	70
60	69	78	40	75	56
70	51	99	69	95	86
57	53	47	50	55	81

   Determine the granularity of this data.

3. A pair of dice, one red and one green, is rolled in the following manner. On the first toss both dice are tossed. Then the red die is tossed while the green die keeps its value to get the second sum. Then the green die is tossed while the red die keeps its value. You continue tossing alternately the red then the green, die and generate the following table.

8	11	8	6	10	9	9
7	2	5	10	11	8	5
3	6	9	7	7	9	10
7	3	3	7	9	8	7
5	4	8	11	11	10	10

Determine the granularity of this data.

2.5 Maximum Number of Groups

THERE ARE TWO formulas used to calculate the maximum number of group. These formulas, when used in conjunction with the minimum number of groups give guidance on how many groups should be used to group the data. Again, we let g be the number of groups and assume there are n observations in the sample.

1. $g \leq \lfloor \sqrt{n+1} \rfloor$ and

2. $g \leq \left\lfloor \dfrac{x_{(n)} - x_{(1)}}{G} \right\rfloor$,

where G is the granularity and the data are assumed to be sorted from smallest to largest.

In the above $\lfloor x \rfloor$ is the **greatest integer function** and means that you should take the largest integer less than or equal to x.

2.5. MAXIMUM NUMBER OF GROUPS

The first formula is based solely on the number of data. The idea is that the more data in each interval the greater the loss in precision.

The second formula considers the granularity. If the width between distinct values is large in comparison to the difference between the largest and smallest datum, then the number of groups calculated merely by the sample size is probably too many and will cause empty intervals to occur. The second formula takes care of this problem. In general, you should use whichever formula produces the fewer number of groups. That is,

$$g \leq \min(\text{formula 1}, \text{formula 2}).$$

If there are more than 15[1] groups then people will find it very difficult to learn anything from the grouping. Therefore, many people add this criterion about the maximum number of groups.

Example 1:
Determine the maximum number of groups need for the oyster data of Example 1, §2.1, which starts on page 20.

Answer:
There are 50 data. The first formula says

$$g \leq \lfloor \sqrt{51} \rfloor \approx \lfloor 7.1414 \rfloor = 7.$$

In Example 1, §2.4, which started on page 33, we determined the granularity is $G = 0.43$. The data was sorted in Example 1, §2.1, starting on page 20, so $x(1) = 5.76$ and $x_{(50)} = 13.18$. We conclude from the granularity that

$$g \leq \left\lfloor \frac{13.18 - 5.76}{0.43} \right\rfloor \approx \lfloor 17.2558 \rfloor = 17.$$

---
[1]Some people make this 11 groups. Your instructor may have his/her own criterion on the maximum number of groups.

Clearly the number of data has more impact than the granularity. We conclude we must have
$$g \le \min(7, 17) = 7;$$
that is, there should be at most 7 groups.

In Example 1, starting on page 30, we determined that there should be at least 4 groups. Therefore, we should use 4, 5, 6 or 7 groups. If you add the odd number of groups restriction then this reduces to 5 or 7 groups.

Example 2:

A certain event occurs about one-third of the time. The table below gives the number of tries needed to get two occurrences of this event. The experiment was repeated 100 times with the following results.

2	11	5	3	5	7	5	6	3	28
4	16	10	14	5	7	12	10	8	3
3	4	3	5	5	2	5	17	2	14
6	2	9	9	4	6	6	6	11	4
2	5	3	8	2	5	4	6	7	10
9	15	12	10	8	14	4	5	5	4
3	11	4	9	5	4	4	5	3	7
14	4	14	8	2	6	4	5	4	6
2	3	3	3	5	9	4	10	3	11
3	3	6	4	3	2	3	5	10	5

Determine the maximum number of groups needed for this data.

Answer:

There are 100 data, so the first formula says there should be at most
$$g \le \lfloor \sqrt{101} \rfloor \approx \lfloor 10.0499 \rfloor = 10.$$

In Example 2, §2.4, starting on page 34, we determined the granularity is 11. The data was also sorted in that example, so $x_{(1)} = 2$ and $x_{(100)} = 28$.
$$g \le \left\lfloor \frac{28 - 2}{11} \right\rfloor = \lfloor 2.36 \rfloor = 2.$$

2.5. MAXIMUM NUMBER OF GROUPS

Many people would regard this as too small. If we use the more reasonable $G = 2$, with a top of 17, then

$$g \le \left\lfloor \frac{17-2}{2} \right\rfloor = \lfloor 7.5 \rfloor = 7.$$

We conclude that at most 7 groups should be used. The granularity is the important feature here, as is the fact that 28 seems unusually far away from the rest of the data.

In Example 2, §2.3, starting on page 30, we determined that there should be at least 5 groups. Therefore, we should use from 5 to 7 groups.

Exercises 2.5

1. One day while walking through a field you notice there seem to be a lot of stones lying on the ground. Curious about what the average size of a stone is in the field you randomly collect 30 stones. You weigh each stone and record its weight in grams. The following table shows the raw data.

110	126	75	110	126	94
90	112	111	139	131	92
102	103	96	75	67	101
104	111	100	65	92	93
106	104	93	115	109	95

   Determine the maximum number of groups needed for this data. Compare your result with the minimum number of groups in Section 2.3.

2. The scores, in the order graded, on a recent midterm examination in Statistics, a class with 30 students, are given below.

70	80	99	98	85	89
87	79	83	38	69	70
60	69	78	40	75	56
70	51	99	69	95	86
57	53	47	50	55	81

Determine the maximum number of groups needed for this data. Compare your result with the minimum number of groups in Section 2.3.

3. A pair of dice, one red and one green, is rolled in the following manner. On the first toss both dice are tossed. Then the red die is tossed while the green die keeps its value to get the second sum. Then the green die is tossed while the red die keeps its value. You continue tossing alternately the red then the green, die and generate the following table.

8	11	8	6	10	9	9
7	2	5	10	11	8	5
3	6	9	7	7	9	10
7	3	3	7	9	8	7
5	4	8	11	11	10	10

Determine the maximum number of groups needed for this data. Compare your result with the minimum number of groups in Section 2.3.

2.6 Creating the Group Intervals

ARMED WITH the minimum and maximum number of groups, it is time to create the group intervals. There are three things that must be kept in mind:

1. The intervals must cover all of the data.

2.6. CREATING THE GROUP INTERVALS

2. The groups will eventually be used in calculations, so the midpoint of each interval should be easy to use.

3. The placement of all data must be certain — no datum can fall equally well into two different intervals.

We will assume that all of the intervals will have the same size. When this is true then a fourth criterion is added:

4. The groups should be centered over the interval spanning the smallest datum to the largest datum.

We will use the four criteria above to help us determine equal-width group intervals. We will let w be this common width and g be the number of intervals desired. We assume the data have been sorted so that $x_{(1)}$ is the smallest value and $x_{(n)}$ is the largest value.

The criterion that the intervals should cover the data means

$$w \geq \frac{x_{(n)} - x_{(1)}}{g}.$$

The criterion that the center points should be easy to use means that the above value of w should be rounded up[1] to some nice value. The nice values are 5, 2 and 1. Usually this means that the values are rounded to the nearest integer (1), half (0.5), fifth (0.2) or tenth (0.1). If more decimal places are needed, then it is rounded up to the nearest 5, 2 or 1 in the last decimal place kept.

The criterion that placement of data should be certain means that endpoints lying between consecutive intervals must belong to exactly one of the touching intervals. By tradition, each endpoint is assigned to the right interval. This means that all intervals have the

---
[1]Thus the $>$ part of the $\geq$ in the formula.

form $[a, b)$, $a \leq x < b$, except for the rightmost interval — it must include its right endpoint.[2]

The endpoints are labeled $t_0, t_1, \ldots, t_g$ and for all $i = 1, \ldots, g$ we must have
$$t_i = t_0 + i \cdot w,$$
where w is the common width. The fourth criterion tells us where to place t_0. After this, the above equation sets the other g values. To be centered over the spanning interval we must have
$$x_{(1)} - t_0 = t_g - x_{(n)}.$$

Expanding the formula for t_g we get
$$x_{(1)} - t_0 = t_0 + g \cdot w - x_{(n)}.$$

Solving for t_0 we get
$$2t_0 = x_{(1)} + x_{(n)} - g \cdot w; \text{ or, } t_0 = \frac{x_{(1)} + x_{(n)} - g \cdot w}{2}.$$

Example 1:
From previous calculations with the oyster data, there should be 4–7 group intervals. Determine some possible intervals for each case.

Answer:
  The fact there is more than one nice ending means there is some freedom in choosing the interval widths; thus, other answers will be possible for each case.

---

[2] Some people make the intervals of the form $(a, b]$, $a < x \leq b$. Then it is the *leftmost* interval which must include its left endpoint. Use whichever form your instructor prefers.

2.6. CREATING THE GROUP INTERVALS

The data were sorted in Example 1, §2.1, starting on page 20. Thus, $x_{(1)} = 5.76$ and $x_{(50)} = 13.18$ and there are 50 data.

For 4 intervals we get a uniform width of

$$w \geq \frac{13.18 - 5.76}{4} = 1.855.$$

All data is two decimal places, so we round this up, always up, to the nearest 0.02 and get $w = 1.86$. The starting value is

$$t_0 = \frac{5.76 + 13.18 - 4(1.86)}{2} = 5.75.$$

This makes the endpoints 5.75, 7.61, 9.47, 11.33 and 13.19. The intervals are therefore $[5.75, 7.61)$, $[7.61, 9.47)$, $[9.47, 11.33)$, and $[11.33, 13.19]$.

For 5 intervals, the uniform width becomes

$$w \geq \frac{13.18 - 5.76}{5} = 1.484.$$

The data are all given to two decimal places and we again choose to round up to the nearest 0.02. This makes $w = 1.50$. The starting value is now

$$t_0 = \frac{5.76 + 13.18 - 5(1.50)}{2} = 5.72.$$

The new endpoints are 5.72, 7.22, 8.72, 10.22, 11.72 and 13.22. The five group intervals have now become $[5.72, 7.22)$, $[7.22, 8.72)$, $[8.72, 10.22)$, $[10.22, 11.72)$ and $[11.72, 13.22]$.

We could have chosen $w = 1.49$, rounding up to the nearest 0.01, but then the midpoints of all intervals would require 3 decimal places. We prefer to make the midpoints match the rest of the data — this is *not* a requirement.

For 6 intervals, we calculate the common width of all intervals to be

$$w \geq \frac{13.18 - 5.76}{6} \approx 1.23667.$$

Rounding to the nearest 0.01 gives $w = 1.24$. With this interval width, the starting point is

$$t_0 = \frac{5.76 + 13.18 - 6(1.24)}{2} = 5.75.$$

The endpoints are 5.75, 6.99, 8.23, 9.47, 10.71, 11.95 and 13.19. The six group intervals are $[5.75, 6.99)$, $[6.99, 8.23)$, $[8.23, 9.47)$, $[9.47, 10.71)$, $[10.71, 11.95)$ and $[11.95, 13.19]$.

Finally, using 7 intervals gives a common width of

$$w = \frac{13.18 - 5.76}{7} = 1.06$$

which is already nice. This makes the starting value 5.76 — check this for yourself! The endpoints are therefore, 5.76, 6.82, 7.88, 8.94, 10.00, 11.06, 12.12 and 13.18. We conclude that the intervals are $[5.76, 6.82)$, $[6.82, 7.88)$, $[7.88, 8.94)$, $[8.94, 10.00)$, $[10.00, 11.06)$, $[11.06, 12.12)$ and $[12.12, 13.18]$.

One feature of the nice endings is that the number of decimal places used for the midpoints of the intervals is never more than one decimal place more than the decimal places used in the endpoints. This makes calculations faster when based upon the midpoints of the intervals — as we will see that they are.

Example 2:
Previous calculations with the two-success data shows that from 5 to 7 intervals are needed. Determine some possible intervals for this data.

Answer:
   In Example 2, §2.3, starting on page 30 we gave the sorted data. There we determined there are 100 data with $x_{(1)} = 2$ and $x_{(100)} = 28$. The

2.6. CREATING THE GROUP INTERVALS

intervals were only calculated to go to 17, we will then add one more to go to 28 that will be larger than the others

If we use 5 intervals then the common width up to 17 will be

$$w = \frac{17-2}{5} = 3.$$

This is nice and the starting value will be 2 for

$$t_0 = \frac{2 + 17 - 5(3)}{2} = 2.$$

The endpoints are 2, 5, 8, 11, 14, 17 and 28 — it was not included in the 5 intervals, so we need an extra large interval to cover it. The six group intervals are $[2,5)$, $[5,8)$, $[8,11)$, $[11,14)$, $[14,17)$ and $[17,28]$.

If we use 6 intervals then the common width up to 17 will be

$$w = \frac{17-2}{6} = 2.5.$$

Again this is nice and the starting value will be 2. The endpoints are 2, 4.5, 7, 9.5, 12, 14.5, 17 and 28 — as we need to cover the last point. The seven group intervals are $[2, 4.5)$, $[4.5, 7)$, $[7, 9.5)$, $[9.5, 12)$, $[12, 14.5)$, $[14.5, 17)$ and $[17, 28]$.

If we use 7 intervals then the common width up to 17 will be

$$w = \frac{17-2}{7} = 2.\overline{142857}$$

We round this up to the nearest 0.1 to get a width of $w = 2.2$. Normally the starting point would be

$$t_0 = \frac{2 + 17 - 7(2.2)}{2} = 1.8;$$

however, $x_{\min} = \mu_{\min}$ and one cannot go below the population minimum — ever!!! [One can never go above the population maximum, either.] We still use t_0 to generate the rest of the endpoints, but the first interval will be a little shorter. Thus, the endpoints are 2, 4, 6.2, 8.4, 10.6, 12.8, 15, 17.2 and 28 — don't forget the last point! This makes the eight group intervals [2, 4), [4, 6.2), [6.2, 8.4), [8.4, 10.6), [10.6, 12.8), [12.8, 15), [15, 17.2) and [17.2, 28].

The problem was that 28 was considerably farther from the other data. Almost half the overall distance from 2 to 28 was in the difference between the ultimate and penultimate points: $28 - 17 = 11$. We have a choice, make them all the same length and accept that some intervals will be empty or create one large interval to cover that huge difference. We chose to do the latter; follow your instructor's advice on what *you* should do.

Remember: One cannot exceed the population limits, so the first and/or the last interval may need to be adjusted. All other endpoints are generated using the formula for $t_0, \ldots, t_g$ as given above.

Exercises 2.6

1. One day while walking through a field you notice there seem to be a lot of stones lying on the ground. Curious about what the average size of a stone is in the field you randomly collect 30 stones. You weigh each stone and record its weight in grams. The following table shows the raw data.

110	126	75	110	126	94
90	112	111	139	131	92
102	103	96	75	67	101
104	111	100	65	92	93
106	104	93	115	109	95

   Use previously calculated results to determine some intervals for all possible cases of group values.

2.6. CREATING THE GROUP INTERVALS

2. The scores, in the order graded, on a recent midterm examination in Statistics, a class with 30 students, are given below.

70	80	99	98	85	89
87	79	83	38	69	70
60	69	78	40	75	56
70	51	99	69	95	86
57	53	47	50	55	81

Use previously calculated results to determine some intervals for all possible cases of group values.

3. A pair of dice, one red and one green, is rolled in the following manner. On the first toss both dice are tossed. Then the red die is tossed while the green die keeps its value to get the second sum. Then the green die is tossed while the red die keeps its value. You continue tossing alternately the red then the green, die and generate the following table.

8	11	8	6	10	9	9
7	2	5	10	11	8	5
3	6	9	7	7	9	10
7	3	3	7	9	8	7
5	4	8	11	11	10	10

Use previously calculated results to determine some intervals for all possible cases of group values.

Chapter 3

Visualizing Data

This chapter begins with frequency tables using both raw and grouped data. It then covers a wide variety of visualizations beginning with point diagrams and ending with Pareto diagrams. Both qualitative and quantitative data are discussed, unless the visualization requires a particular type of data.

3.1 Frequency Tables

THE SIMPLEST OF all visualizations is hardly a visualization at all. The **frequency table** or **tally table** is a table which lists each distinct observation together with the number of times that observation appears in the sample. Frequency tables can be done with grouped or ungrouped data and work for both qualitative and quantitative data. Although not required, frequency tables are easier to construct with sorted data — especially in the grouped version.

Ungrouped Data

THE DATA DOES not need to be grouped, although if there are many distinct values then it probably should be. We first consider ungrouped data.

The **frequency** of an observation x, denoted **frq(x)**, is the number of times x is observed in the sample. A **frequency table** is a listing of the distinct observations and their frequencies.

Example 1:
Determine a frequency table for the two-success data. The sorted data are presented below for convenience.

2	2	2	2	2	2	2	2	2	3
3	3	3	3	3	3	3	3	3	3
3	3	3	3	3	4	4	4	4	4
4	4	4	4	4	4	4	4	4	4
5	5	5	5	5	5	5	5	5	5
5	5	5	5	5	5	5	6	6	6
6	6	6	6	6	6	7	7	7	7
8	8	8	8	9	9	9	9	9	10
10	10	10	10	10	11	11	11	11	12
12	14	14	14	14	14	15	16	17	28

Answer:
From the sorted data there are sixteen distinct values. The frequency table for the data is

2: 9	4: 15	6: 9	8: 4	10: 6	12: 2	15: 1	17: 1
3: 16	5: 17	7: 4	9: 5	11: 4	14: 5	16: 1	28: 1

Observe how the frequency table is listed in column order rather than row order. This is traditional as it makes the variations of the frequency table easier to construct.

3.1. FREQUENCY TABLES

There is no need for the data to be quantitative, qualitative data can be counted just as easily.

Example 2:
Create a frequency table for the fuþorc data of Example 2, §2.1, starting on page 22. The sorted data are repeated below.

ᚠ	ᚠ	ᚢ	ᚢ	ᚢ	ᚦ	ᚦ	ᚦ	ᚩ	ᚩ	ᚩ	ᚱ	ᚱ	ᚱ	ᚱ
ᚱ	ᚱ	ᚳ	ᚳ	ᚷ	ᚷ	ᚹ	ᚹ	ᚻ	ᚻ	ᚻ	ᚻ	ᚾ	ᚾ	ᚾ
ᚾ	ᚾ	ᚾ	ᛁ	ᛁ	ᛁ	ᛁ	ᛁ	ᛁ	ᛄ	ᛄ	ᛄ	ᛄ	ᛄ	ᛇ
ᛈ	ᛈ	ᛉ	ᛉ	ᛉ	ᛋ	ᛋ	ᛋ	ᛋ	ᛏ	ᛏ	ᛏ	ᛏ	ᛒ	ᛒ
ᛒ	ᛒ	ᛗ	ᛗ	ᛗ	ᛗ	ᛗ	ᛚ	ᛚ	ᛝ	ᛝ	ᛝ	ᛝ	ᛟ	ᛟ
ᛟ	ᛟ	ᛟ	ᛟ	ᛟ	ᛞ	ᛞ	ᛞ	ᛞ	ᚪ	ᚪ	ᚪ	ᚪ	ᚫ
ᚫ	ᚣ	ᚣ	ᛡ	ᛡ	ᛡ	ᛠ	ᛠ	ᛣ	ᛣ	ᛣ	ᛤ	ᛤ	ᛤ

Answer:
There are 33 letters in the fuþorc. The frequency table is listed below.

ᚠ:2	ᚳ:2	ᛁ:6	ᛋ:4	ᛚ:2	ᚫ:1	ᛠ:2
ᚢ:3	ᚷ:2	ᛄ:5	ᛏ:4	ᛝ:4	ᛞ:4	ᛣ:4
ᚦ:3	ᚹ:2	ᛇ:1	ᛒ:4	ᛟ:2	ᛡ:2	ᛤ:3
ᚩ:3	ᚻ:4	ᛈ:2	ᛗ:1	ᚪ:6	ᚣ:2
ᚱ:6	ᚾ:6	ᛉ:3	ᛗ:4	ᚫ:3	ᛠ:3

The order of the fuþorc can be read reading down each column and skipping the frequency columns.

Example 3:
Give a frequency table for the balloon data of Example 3, §2.1, starting on page 23. The data, sorted in alphabetic order, are

blue	blue	blue	blue	blue
blue	green	green	green	green
orange	orange	orange	orange	orange
red	red	red	red	red
yellow	yellow	yellow	yellow	yellow

Answer:
> There are five colors. The frequency table is
>
> blue : 6   orange: 5   yellow: 5
> green: 4   red   : 5

Example 4:
Display a frequency table for the candy-coated chocolate drops data of Example 4, §2.1, starting on page 24. The data sorted in alphabetic order are listed below.

blue	blue	blue	blue	blue	blue	blue	blue
blue	brown	brown	brown	brown	brown	brown	brown
brown	brown	brown	brown	brown	brown	green	green
green	orange	orange	orange	orange	orange	orange	orange
orange	red	red	red	red	red	red	red
red	yellow	yellow	yellow	yellow	yellow	yellow	yellow

Answer:
> There are six colors for the candy coating. The frequency table is
>
> blue : 9   green : 3   red   : 8
> brown: 13   orange: 8   yellow: 7

Grouped Data

THERE ISN'T A LARGE difference between grouped and ungrouped data. The frequency of an interval is the sum of the frequencies for the distinct observations that fall into the interval. One other distinction is that the midpoint of the interval is normally also listed.

Example 5:
Create a frequency table for the oyster data when the data are placed

3.1. FREQUENCY TABLES

into 5 group intervals. Use the intervals from Example 1, §2.6, starting on page 42 in the 5 interval category.

Answer:
Using intervals $[5.72, 7.22)$, $[7.22, 8.72)$, $[8.72, 10.22)$, $[10.22, 11.72)$ and $[11.72, 13.22]$. The format we use below is common when the number of intervals is small.

Interval	$[5.72, 7.22)$	$[7.22, 8.72)$	$[8.72, 10.22)$	$[10.22, 11.72)$	$[11.72, 13.22]$
Midpoint	6.47	7.97	9.47	10.97	12.47
Frequency	14	12	8	13	3

Example 6:
Repeat Example 5 above with the data divided into seven groups.

Answer:
The intervals used are from the same example, but under the category 7 intervals. This frequency table shows the other format.

Interval	Midpoint	Frequency
$[5.76, 6.82)$	6.29	9
$[6.82, 7.88)$	7.35	8
$[7.88, 8.94)$	8.41	9
$[8.94, 10.00)$	9.47	8
$[10.00, 11.06)$	10.53	9
$[11.06, 12.12)$	11.59	6
$[12.12, 13.18]$	12.65	1

Exercises 3.1

1. A bag of candy-coated chocolate drops has 42 pieces of candy in it. You recorded the colors of each piece as it was removed from the bag

and the result is given in the follow table.

red	brown	yellow	brown	green	yellow	yellow
brown	brown	brown	brown	red	brown	red
brown	brown	blue	brown	yellow	brown	blue
green	brown	yellow	yellow	brown	brown	brown
yellow	orange	orange	yellow	yellow	red	brown
brown	blue	red	blue	yellow	yellow	red

Sort these data alphabetically by the name of the color and create a frequency table from the result.

2. You toss a coin 20 times and record whether the obverse side (heads) or reverse side (tails) fell up. The result of these twenty tosses is shown below. In the table, H means heads and T means tails.

T	H	T	T	H
H	T	T	H	T
H	H	T	T	H
T	H	H	T	H

Create a frequency table for the result.

3. One day while walking through a field you notice there seem to be a lot of stones lying on the ground. Curious about what the average size of a stone is in the field you randomly collect 30 stones. You weigh each stone and record its weight in grams. The following table shows the raw data.

110	126	75	110	126	94
90	112	111	139	131	92
102	103	96	75	67	101
104	111	100	65	92	93
106	104	93	115	109	95

Create a frequency data for the ungrouped data. Repeat for grouped data using one of the numbers of possible groups derived earlier.

4. The scores, in the order graded, on a recent midterm examination in Statistics, a class with 30 students, are given below.

3.2. FREQUENCY TABLE VARIATIONS

70	80	99	98	85	89
87	79	83	38	69	70
60	69	78	40	75	56
70	51	99	69	95	86
57	53	47	50	55	81

Create a frequency data for the ungrouped data. Repeat for grouped data using one of the numbers of possible groups derived earlier.

5. A pair of dice, one red and one green, is rolled in the following manner. On the first toss both dice are tossed. Then the red die is tossed while the green die keeps its value to get the second sum. Then the green die is tossed while the red die keeps its value. You continue tossing alternately the red then the green, die and generate the following table.

8	11	8	6	10	9	9
7	2	5	10	11	8	5
3	6	9	7	7	9	10
7	3	3	7	9	8	7
5	4	8	11	11	10	10

Create a frequency data for the ungrouped data. Repeat for grouped data using one of the numbers of possible groups derived earlier.

3.2 Frequency Table Variations

THERE ARE A FEW common variations for the basic frequency table. These variations will be discussed here.

Probability Tables

THE FIRST VARIATION is the **probability table** or **relative frequency** table. Each frequency is divided by the total number of observations in the sample to get the **empirical probability**. This can be done with either qualitative or quantitative data. These are also called **point distributions** or **point probability distributions**.

Example 1:
The letters of the fuþorc can be divided into simple consonants, multiple consonants, vowels and semivowels according to the following chart:

Simple Consonant (SC)	ᚱ·ᚴ·ᚲ·ᛏ·ᛒ·ᛗ·ᚾ·ᚷ·ᛊ·ᛚ·ᛜ·ᚺ
Multiple Consonant (MC)	ᚠ·ᚦ·ᚳ·ᛉ·ᛣ·ᚻ·ᛝ
Vowel (V)	ᚢ·ᚩ·ᛁ·ᛄ·ᛘ·ᛟ·ᛖ·ᛡ·ᛠ
Semivowel (SV)	ᛈ·ᚻ·ᛟ·ᛙ·ᛯ

A multiple consonant is one which represents more than one sound. The letter ᚠ *is used for both the allophone v and the fricative f. The letter* ᚦ *is used for both the allophone þ and the fricative ð.*

Create a point distribution for the data.

Answer:
We first create a frequency table based upon these categories. The frequencies are the sum of the frequencies for the letters in each category.

Category	SC	MC	V	SV
Frequency	44	18	26	17

Dividing each of these by 105, the number of observations, and rounding each value to 3 decimal places gives

Category	SC	MC	V	SV
Frequency	0.419	0.171	0.248	0.162

One often finds that the sum will be slightly more or slightly less than 1 because each entry is rounded separately. Here we "lucked

3.2. FREQUENCY TABLE VARIATIONS

out" two rounded up and two rounded down.

Example 2:
Determine a probability table for the balloon data.

Answer:
There were 25 balloons drawn, so these probabilities will add up to 1.00 as $1/25 = 0.04$. The probability table is given below.

Color	blue	green	orange	red	yellow
Pr(Color)	0.24	0.16	0.20	0.20	0.20

In the above, no rounding was necessary.

Numeric data can be used in the probability table, as well. Normally, this is used as grouped data.

Example 3:
Create a probability table for the two-success data with six intervals.

Answer:
The sorted data are presented below for convenience.

2	2	2	2	2	2	2	2	2	3
3	3	3	3	3	3	3	3	3	3
3	3	3	3	3	4	4	4	4	4
4	4	4	4	4	4	4	4	4	4
5	5	5	5	5	5	5	5	5	5
5	5	5	5	5	5	5	6	6	6
6	6	6	6	6	6	7	7	7	7
8	8	8	8	9	9	9	9	9	10
10	10	10	10	10	11	11	11	11	12
12	14	14	14	14	14	15	16	17	28

There were six intervals determined in Example 2, §2.6, starting on page 44. They are $[2, 5)$, $[5, 8)$, $[8, 11)$, $[11, 14)$, $[14, 17)$ and $[17, 28]$. Recall: Frequency tables, and therefore probability tables, often list the midpoint information for each interval, as well. Below we show a combined frequency/probability table.

Interval (I)	[2, 5)	[5, 8)	[8, 11)	[11, 14)	[14, 17)	[17, 28]
Midpoint	3.5	6.5	9.5	12.5	15.5	22.5
Frequency	40	30	15	6	7	2
$\Pr(X \in I)$	0.40	0.30	0.15	0.06	0.07	0.02

The other choice is to create six intervals spanning the entire set. The common width is

$$w = \frac{28 - 2}{6} \approx 4.333$$

We will round this up to the nearest 0.1 to get $w = 4.4$. The starting point for generating endpoints is

$$t_0 = \frac{2 + 28 - 6(4.4)}{2} = 1.8,$$

but we know that we cannot go lower than 2, as this is the two-successes data, so $\mu_{\min} = 2$. The endpoints are 2, 6.2, 10.6, 15, 19.4, 23.8, 28.2. The group intervals become [2, 6.2), [6.2, 10.6), [10.6, 15), [15, 19.4), [19.4, 23.8), [23.8, 28.2]. Which gives the following probability distribution.

Interval (I)	[2, 6.2)	[6.2, 10.6)	[10.6, 15)	[15, 19.4)	[19.4, 23.8)	[23.8, 28.2]
Midpoint	4.1	8.4	12.8	17.2	21.6	26.0
Frequency	66	19	11	3	0	1
$\Pr(X \in I)$	0.66	0.19	0.11	0.03	0.00	0.01

There is nothing in the interval [19.4, 23.8). This is what we meant about there being an empty interval. the nice thing about empty intervals is they clearly show when items are far separated from the normal run of data.

Cumulative Tables

THE SECOND VARIATION of a frequency table is used for numeric data only. This can be used with both grouped and ungrouped data,

3.2. FREQUENCY TABLE VARIATIONS

but it is far more common to do it with grouped data. A **cumulative table**, **cumulative frequency table** or **cumulative tally table** has the sum of all frequencies from the smallest value up to and including the current distinct observation or interval. If grouped data is used, it will often have the midpoint of each interval listed.

Example 4:
Show the cumulative table for the oyster data grouped into seven intervals.

Answer:
   Below is the table, a simple modification of the table given in Example 6, §3.1, starting on page 53.

Interval	Midpoint	Cumulation
[5.76, 6.82)	6.29	9
[6.82, 7.88)	7.35	17
[7.88, 8.94)	8.41	26
[8.94, 10.00)	9.47	34
[10.00, 11.06)	10.53	43
[11.06, 12.12)	11.59	49
[12.12, 13.18]	12.65	50

Cumulative Distributions

IN THE FINAL variation, instead of the correspondent being the cumulative frequency, that is now divided by the total number of observations in the sample. These are best created directly from the cumulative table, to ensure the final interval pairs to 1. This means the difference between consecutive entries, when multiplied by the total number of observations, may not give an exact integer.

Example 5:
Create the cumulative distribution for the oyster data grouped into seven intervals.

Answer:
   Below is the table, a simple modification of the table given in Example 4 above.

Interval	Midpoint	Cumulation
[5.76, 6.82)	6.29	0.18
[6.82, 7.88)	7.35	0.34
[7.88, 8.94)	8.41	0.52
[8.94, 10.00)	9.47	0.68
[10.00, 11.06)	10.53	0.86
[11.06, 12.12)	11.59	0.98
[12.12, 13.18]	12.65	1.00

Exercises 3.2

1. A bag of candy-coated chocolate drops has 42 pieces of candy in it. You recorded the colors of each piece as it was removed from the bag and the result is given in the follow table.

red	brown	yellow	brown	green	yellow	yellow
brown	brown	brown	brown	red	brown	red
brown	brown	blue	brown	yellow	brown	blue
green	brown	yellow	yellow	brown	brown	brown
yellow	orange	orange	yellow	yellow	red	brown
brown	blue	red	blue	yellow	yellow	red

Sort these data alphabetically by the name of the color and create a probability table from the result.

3.2. FREQUENCY TABLE VARIATIONS

2. You toss a coin 20 times and record whether the obverse side (heads) or reverse side (tails) fell up. The result of these twenty tosses is shown below. In the table, H means heads and T means tails.

T	H	T	T	H
H	T	T	H	T
H	H	T	T	H
T	H	H	T	H

Create a probability table for the result.

3. One day while walking through a field you notice there seem to be a lot of stones lying on the ground. Curious about what the average size of a stone is in the field you randomly collect 30 stones. You weigh each stone and record its weight in grams. The following table shows the raw data.

110	126	75	110	126	94
90	112	111	139	131	92
102	103	96	75	67	101
104	111	100	65	92	93
106	104	93	115	109	95

Create a probability data for the ungrouped data. Then create a cumulative table and cumulative distribution using one of the possible grouping values derived earlier.

4. The scores, in the order graded, on a recent midterm examination in Statistics, a class with 30 students, are given below.

70	80	99	98	85	89
87	79	83	38	69	70
60	69	78	40	75	56
70	51	99	69	95	86
57	53	47	50	55	81

Create a probability data for the ungrouped data. Then create a cumulative table and cumulative distribution using one of the possible grouping values derived earlier.

5. A pair of dice, one red and one green, is rolled in the following manner. On the first toss both dice are tossed. Then the red die is tossed while the green die keeps its value to get the second sum. Then the green die is tossed while the red die keeps its value. You continue tossing alternately the red then the green, die and generate the following table.

8	11	8	6	10	9	9
7	2	5	10	11	8	5
3	6	9	7	7	9	10
7	3	3	7	9	8	7
5	4	8	11	11	10	10

Create a probability data for the ungrouped data. Then create a cumulative table and cumulative distribution using one of the possible grouping values derived earlier.

3.3 Point Diagrams

THIS IS WHERE the visualizations truly begin. The **point diagram** or **point plot** is a visual representation of a frequency diagram. Point plots are primarily used for ungrouped, quantitative data when there are only a few distinct values. A generalization of the point diagram for grouped quantitative data will be covered in different section. We will however show a point plot applied to qualitative data, but the visualization normally used for qualitative data will be covered in a different section.

When used for quantitative data, the observed values are plotted along a number line. The number line can be horizontal or vertical. If the number line is horizontal then a tower of Xs is shown above the point, one X for each occurrence of the observation. Many people

3.3. POINT DIAGRAMS

will evenly space the observed values instead of plotting them along a number line. This is especially true when the difference between the smallest and largest values is large compared to the difference between some of the consecutive observed values.

Example 1:
The following numbers represent a random sample drawn of 25 data from a gamma distribution and rounded to the nearest integer.

12	5	9	24	4
13	5	27	18	9
13	14	5	8	24
17	22	17	13	31
5	10	15	14	22

Create a point diagram for this data. Show two versions. The first plotting observed values along a number line. The second using equal spacing.

Answer:
We first sort the data to get

4	5	5	5	5
8	9	9	10	12
13	13	13	14	14
15	17	17	18	22
22	24	24	27	31

The frequency table for this data is:

4: 1	9: 2	13: 3	17: 2	24: 2
5: 4	10: 1	14: 2	18: 1	27: 1
8: 1	12: 1	15: 1	22: 2	31: 1

When the values are plotted along a number line we get the following point diagram.

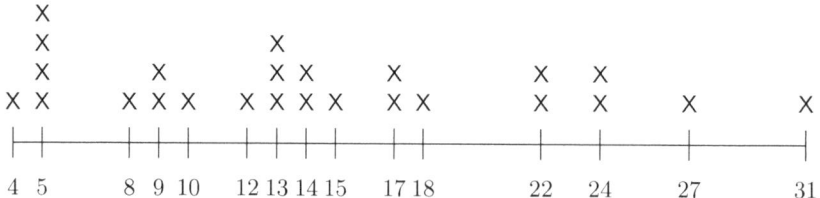

When the values are equally spaced the point plot becomes as shown below.

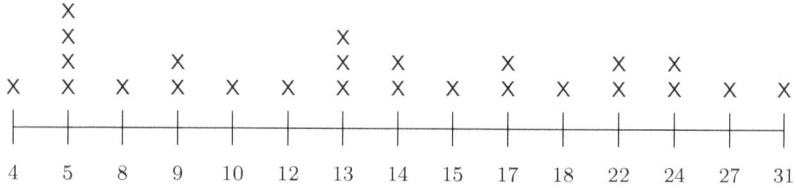

What gets lost when you equally space them is the existence of large gaps between several of the consecutive values. Yes, you can get that from the numbers, but it doesn't have the visual impact that plotting them along the number line has.

If the data are qualitative then there is no number line over which you can plot them. In this case they may as well be equally spaced. If the line along which they are placed is vertical then a row of Xs is used. Again, one for each occurrence. This is best when the frequencies are fairly large and plotting them vertically would waste space.

Example 2:

Create a point plot for the candy-coated chocolate drops.

Answer:

3.3. POINT DIAGRAMS

A frequency table of the data is shown below.

$$\begin{array}{lll} \text{blue}: 9 & \text{green}: 3 & \text{red}: 8 \\ \text{brown}: 13 & \text{orange}: 8 & \text{yellow}: 7 \end{array}$$

From the above we get the following point diagram.

```
blue   ┬ X X X X X X X X X
brown  ┼ X X X X X X X X X X X X X
green  ┼ X X X
orange ┼ X X X X X X X X
red    ┼ X X X X X X X X
yellow ┴ X X X X X X X
```

Point diagrams are normally used with numeric data, but as you can see, they can also be used with categorical data.

Exercises 3.3

1. A bag of candy-coated chocolate drops has 42 pieces of candy in it. You recorded the colors of each piece as it was removed from the bag and the result is given in the follow table.

red	brown	yellow	brown	green	yellow	yellow
brown	brown	brown	brown	red	brown	red
brown	brown	blue	brown	yellow	brown	blue
green	brown	yellow	yellow	brown	brown	brown
yellow	orange	orange	yellow	yellow	red	brown
brown	blue	red	blue	yellow	yellow	red

    Create a point diagram for this data.

2. You toss a coin 20 times and record whether the obverse side (heads) or reverse side (tails) fell up. The result of these twenty tosses is shown below. In the table, H means heads and T means tails.

T	H	T	T	H
H	T	T	H	T
H	H	T	T	H
T	H	H	T	H

Create a point plot for the data.

3. A pair of dice, one red and one green, is rolled in the following manner. On the first toss both dice are tossed. Then the red die is tossed while the green die keeps its value to get the second sum. Then the green die is tossed while the red die keeps its value. You continue tossing alternately the red then the green, die and generate the following table.

8	11	8	6	10	9	9
7	2	5	10	11	8	5
3	6	9	7	7	9	10
7	3	3	7	9	8	7
5	4	8	11	11	10	10

Create a point diagram for the data.

3.4 Stem-and-Leaf Diagrams

THERE IS ANOTHER way of using the data when a point diagram may have too many points. This next diagram is most useful when the difference between consecutive differences is relatively small. A **stem-and-leaf diagram** or **stem-and-leaf plot** divides the data into two portions; hence, the data must have at least two digits. The foremost digit(s) is(are) the **stem(s)**. The rearmost digit(s) is(are) the **leaf(leaves)**. A **zero stem** or **null stem** can be used if a few data are one-digit. All leaves must have the same number of digits. The decimal point must be in the same location relative to all

3.4. STEM-AND-LEAF DIAGRAMS

leaves. If there are too many leaves for each stem to comfortably be displayed, then stems may be repeated. All stems should be repeated the same number of times, even those which do not have too many leaves.

At the top of the stem-and-leaf plot are the total number of data and the **scale** of the final digit. The scale is used to locate the decimal point as both stem and leaf are treated as integers.

When the leaf is more than one digit then commas are placed between consecutive leaves. This is not necessary for single digit leaves.

On each side of the stem-and-leaf display there are running totals. The left column counts the total number of data starting from the top and going left-to-right through the current line. The right column counts the total number of data starting from the bottom and going right-to-left through the current line. There is a quick check which can be used to verify the counts are correct: For each row, the sum of the left and right columns, minus the number of data, gives the number of leaves in the that row.

Example 1:
Create a stem-and-leaf diagram for random sample of Example 1, § 3.3, which started on page 63.

Answer:
  The sorted data are repeated here for convenience.

4	5	5	5	5
8	9	9	10	12
13	13	13	14	14
15	17	17	18	22
22	24	24	27	31

The tens-digit is the stem and the ones-digit is the leaf. As suggested, a zero stem is used for the one-digit values. The scale shows that the data are integers with the leaf in the unit position.

	$n = 25$	scale $= 1$	
(8)	0	45555899	(25)
(19)	1	02333445778	(17)
(24)	2	22447	(6)
(25)	3	1	(1)

The original data are easily recovered from a stem-and-leaf display. This is exceptionally easy when the scale is 1, for all one needs to do is attach the leaf onto the stem. If the stem is 0, then just ignore the result. Hence, the smallest datum is $04 = 4$ and the largest is 31. In generally, the scale is used to determine the decimal point and we will see how to do this shortly.

Stem-and-leaf diagrams are also useful to show patterns which may not be obvious from the sorted data or even from the frequency table. For example, the above shows that there are more data in the decade from 10 to 19 than in any other decade. Also, it is clear that smaller values are favored over larger ones. This will give us things to test when we discuss statistics based upon these data.

Example 2:
Create a stem-and-leaf plot for the oyster data.

Answer:
 We will use the last two digits as the leaf. This will make the scale 0.01 to indicate that both digits of the leaf lie to the right of the decimal point. Do not forget the leading zero in the leaf!!! We will need to separate leaves

3.4. STEM-AND-LEAF DIAGRAMS

by a comma as there is more than one digit. The stem-and-leaf diagram is shown below.

	$n = 50$	scale $= 0.01$	
(1)	5	76	(50)
(12)	6	10,36,37,42,53,63,67,74,91,91,97	(49)
(18)	7	03,04,37,40,71,97	(38)
(26)	8	02,04,12,14,42,53,61,62	(32)
(34)	9	01,20,21,39,47,53,55,98	(24)
(42)	10	24,50,62,63,63,81,81,86	(16)
(48)	11	02,09,17,31,59,97	(8)
(49)	12	07	(2)
(50)	13	18	(1)

It is easy to recover the data values using the scale: The lowest observation is $576 \cdot 0.01 = 5.76$. The highest is $1318 \cdot 0.01 = 13.18$. Both of these are easily verified from looking at the sorted data.

The data appears to favor smaller values more than higher values. Averaging over multiple rows, might make the data balanced. Later we will use statistics to decide how balanced these data are.

Example 3:
A certain call center answers the questions of people who phone into it. It receives many calls each day. Below we see the waiting time in seconds between consecutive calls on one particular afternoon. The first time listed is the time between the first call received that afternoon and the second call. The gathered order shows the first fifty waiting times that afternoon.

36	39	44	26	66	48	30	27	43	42
72	59	26	38	24	45	35	29	84	46
49	40	37	24	24	32	38	37	29	88
112	48	50	29	30	75	24	76	47	27
91	71	66	42	43	45	33	35	31	79

First sort the data. Then create a stem-and-leaf diagram for this data.

Answer:

We first sort the data to get

24	24	24	24	26	26	27	27	29	29
29	30	30	31	32	33	35	35	36	37
37	38	38	39	40	42	42	43	43	44
45	45	46	47	48	48	49	50	59	66
66	71	72	75	76	79	84	88	91	112

Using the last digit as the leaf, we get the following stem-and-leaf plot.

	$n = 50$	scale = 1		
(11)	2	44446677999	(50)
(24)	3	0012355677889	(39)
(37)	4	0223345567889	(26)
(39)	5	09	(13)
(41)	6	66	(11)
(46)	7	12569	(9)
(48)	8	48	(4)
(49)	9	1	(2)
(49)	10		(1)
(50)	11	2	(1)

Never skip stems! The fact that stem 10 has no data is significant. We will see this later when we discuss statistics based upon these data.

The dip in the 5–6 score is probably caused by there being more groups than recommended.

The gap between the 91 and the 112 looks like it probably should be there, but will probably get masked with fewer intervals.

3.4. STEM-AND-LEAF DIAGRAMS

The data are heavily favoring smaller values — almost half of the data is in the first two decades!!! This will not change with fewer intervals — it can only get worse. In many respects this resembles the random gamma data.

Exercises 3.4

1. One day while walking through a field you notice there seem to be a lot of stones lying on the ground. Curious about what the average size of a stone is in the field you randomly collect 30 stones. You weigh each stone and record its weight in grams. The following table shows the raw data.

110	126	75	110	126	94
90	112	111	139	131	92
102	103	96	75	67	101
104	111	100	65	92	93
106	104	93	115	109	95

    Create a stem-and-leaf diagram for the data. [Hint: use the final digit as the leaf.]

2. The scores, in the order graded, on a recent midterm examination in Statistics, a class with 30 students, are given below.

70	80	99	98	85	89
87	79	83	38	69	70
60	69	78	40	75	56
70	51	99	69	95	86
57	53	47	50	55	81

    Create a stem-and-leaf plot of the data.

3. Below is a sample of the monthly salary for 30 randomly selected people in mid-management in thousands of dollars.

3.08	3.18	3.13	7.41	3.73	3.05
4.17	3.58	3.36	3.27	3.36	4.74
3.32	3.61	4.26	3.02	3.45	4.06
6.73	4.72	3.13	3.15	3.70	3.59
3.12	3.61	5.03	3.53	3.20	3.32

Create a stem-and-leaf diagram for this data. [Hint: Use the final two digits for the leaf with a scale of 0.01. You might want to repeat each stem twice using leaves 00–49 on the first occurrence and 50–99 on the second. Repeat the stems even if there are only leaves for one of them, leaving the other stem with nothing following it.]

3.5 Bar Charts

THIS SECTION'S TOPIC is normally used with categorical data. There is a variation of this coming up which is normally used with numeric data, but that will be a topic in another section.

The **bar chart** or **bar graph** uses rectangles whose length is proportional to the frequency of the category. The bars can be placed vertically or horizontally and are labeled at one end with the name of the category and often on the other end with the frequency or relative frequency (sometimes as a percentage). If the bar is big enough then the frequency or relative frequency may be placed inside the bar. The bars may be colored or have fill patterns, but normally this is not done unless the categories are not placed next to the bar. In that case a legend is used to identify which category is represented by each bar. Often a title is given to the bar graph which can be placed either above or below.

3.5. BAR CHARTS

Example 1:
Draw a bar graph for the balloon data. The frequency table is given here for convenience.

blue: 6   green: 4   orange: 5   red: 5   yellow: 5

Answer:
The bar chart is shown below.

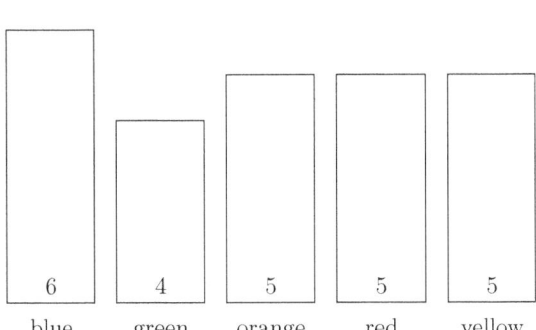

Example 2:
Create a bar chart for the runes as categorized in Example 1, §3.2, starting on page 56.

Answer:
The frequency table is listed below.

Category	SC	MC	V	SV
Frequency	44	18	26	17

This gives the following bar chart:

Random Anglo-Saxon Runes

Legend
SC = Single Consonant
MC = Multiple Consonant
V = Vowel
SV = Semivowel

SC 44
MC 18
V   26
SV 17

Example 3:
A coin is tossed 40 times with the following results.

H	H	T	H	T	T	H	T	H	H
T	T	H	T	H	H	H	H	H	T
H	H	H	H	H	T	H	H	T	H
H	H	T	T	T	H	H	T	H	H

Draw a bar chart for this data. Do you think this is a fair coin?

Answer:

We first make a frequency table — since there are only two outcomes there is no need to sort them to do this. There are 26 H and 14 T which gives the frequency table below

H: 26   T: 14

The bar graph is shown below:

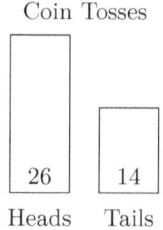

Coin Tosses

26   14
Heads Tails

Exercises 3.5

1. A bag of candy-coated chocolate drops has 42 pieces of candy in it. You recorded the colors of each piece as it was removed from the bag and the result is given in the follow table.

red	brown	yellow	brown	green	yellow	yellow
brown	brown	brown	brown	red	brown	red
brown	brown	blue	brown	yellow	brown	blue
green	brown	yellow	yellow	brown	brown	brown
yellow	orange	orange	yellow	yellow	red	brown
brown	blue	red	blue	yellow	yellow	red

   Create a bar chart of this data.

2. You toss a coin 20 times and record whether the obverse side (heads) or reverse side (tails) fell up. The result of these twenty tosses is shown below. In the table, H means heads and T means tails.

T	H	T	T	H
H	T	T	H	T
H	H	T	T	H
T	H	H	T	H

   Display a bar graph for the tosses.

3. A pair of dice, one red and one green, is rolled in the following manner. On the first toss both dice are tossed. Then the red die is tossed while the green die keeps its value to get the second sum. Then the green die is tossed while the red die keeps its value. You continue tossing alternately the red then the green, die and generate the following table.

8	11	8	6	10	9	9
7	2	5	10	11	8	5
3	6	9	7	7	9	10
7	3	3	7	9	8	7
5	4	8	11	11	10	10

Determine a bar chart for the data.

4. The scores, in the order graded, on a recent midterm examination in Statistics, a class with 30 students, are given below.

$$\begin{array}{cccccc}
70 & 80 & 99 & 98 & 85 & 89 \\
87 & 79 & 83 & 38 & 69 & 70 \\
60 & 69 & 78 & 40 & 75 & 56 \\
70 & 51 & 99 & 69 & 95 & 86 \\
57 & 53 & 47 & 50 & 55 & 81
\end{array}$$

Use the grading scale below to convert the data to nominal form and create a bar graph of the result.

Score at Least	Grade
90	A
80	B
70	C
60	D
0	F

International students may prefer the following scale

Score at Least	Grade
85	Excellent
70	Good
60	Acceptable
50	Poor
0	Fail

3.6 Pie Charts

THIS VISUALIZATION is also used for categorical data. Like a bar chart, it can be used for discrete data, but rarely is. A **pie chart** or **circular diagram** replaces te bars of a bar chart with a circle cut

3.6. PIE CHARTS

into pieces. The area of each sector is proportional to the relative frequency of the category it represents.

The area of a sector is

$$A = \frac{r^2\theta}{2},$$

where r is the radius of the circle and θ is the central angle in radians. One full circle corresponds to an angle of 2π radians or equivalently $360°$. We can use this to determine the central angle for each sector. The area of a circle is πr^2. If the frequency of category i is f_i and the total number of observations in the sample is n, the area of the sector is

$$\frac{f_i \cdot \pi r^2}{n} = \frac{r^2\theta}{2}.$$

Solving for θ we get

$$\theta = \frac{f_i \cdot 2\pi}{n} = \frac{360 f_i}{n}.$$

Pie charts almost always use colors or fill patterns to distinguish the sectors and have a legend to explain what color represents each category. Some people get fancy with 3-dimensional pie graphs, or exploded pie graphs, but here we only cover the basic pie chart. There are three commonly used methods used to fill in the sectors of the pie chart

1. Sort the categories by size, largest first. Then the first sector should straddle the positive x-axis and subsequent sectors are added alternating counterclockwise and clockwise in descending order of size. In this way the smallest sector is "opposite" the largest sector.

2. Sort the categories any way your little heart desires. Place the first category with its starting edge on the positive x-axis and build each sector counterclockwise around the circle.

3. Sort the categories any way you like. The first category has its starting edge on the positive y-axis and each sector is built clockwise around the circle.

Note: When actually calculating the angle, especially on methods 2 and 3, one will end up with better results if one keeps track of the cumulative frequency and calculates differences rather than calculating the angle of each sector directly. Otherwise, the last sector is likely not to be the correct size.

Example 1:
Create a pie chart for the runes as categorized in Example 1, §3.2, starting on page 56.

Answer:
The frequency table is listed below.

Category	SC	MC	V	SV
Frequency	44	18	26	17

We first calculate the angles for each sector:

$$SC \quad \frac{44 \cdot 360}{105} \approx 150.86°$$
$$MC \quad \frac{18 \cdot 360}{105} \approx 61.71°$$
$$V \quad \frac{26 \cdot 360}{105} \approx 89.14°$$
$$SV \quad \frac{17 \cdot 360}{105} \approx 58.29°$$

The angles happen to round correctly to add up to 360°. We used method 2 to create the pie chart in the given order of categories.

3.6. PIE CHARTS

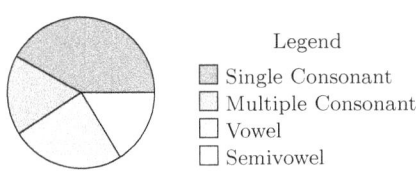

Example 2:
Create a pie graph for the candy-coated chocolate drops data.

Answer:
A frequency table for the data is given below.

blue	: 9	green	: 3	red	: 8
brown:	13	orange:	8	yellow:	7

In decreasing order of frequency, the colors are brown, blue, orange and red are tied, yellow and finally green. We will break the tie by using alphabetic order, so orange will be third and red fourth.

The angles for each sector are

$$\text{brown} \quad \tfrac{13 \cdot 360}{48} = 97.5° \qquad \text{red} \quad \tfrac{8 \cdot 360}{48} = 60.0°$$

$$\text{blue} \quad \tfrac{9 \cdot 360}{48} = 67.5° \qquad \text{yellow} \quad \tfrac{7 \cdot 360}{48} = 52.5°$$

$$\text{orange} \quad \tfrac{8 \cdot 360}{48} = 60.0° \qquad \text{green} \quad \tfrac{3 \cdot 360}{48} = 22.5°$$

The pie chart using method 1 appears below.

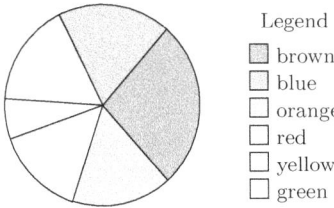

The colors red and orange could have swapped places as these categories had the same frequency.

Example 3:
Make a pie chart for the coin data.

Answer:
The tally table for this data is

$$H: 26 \quad T: 14$$

The angles are

$$H \quad \frac{26 \cdot 360}{40} = 234° \quad T \quad \frac{14 \cdot 360}{40} = 126°$$

Using method 2 and starting with Heads we get the following pie graph.

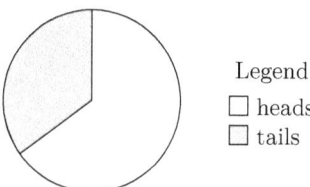

Legend
☐ heads
☐ tails

Exercises 3.6

1. A bag of candy-coated chocolate drops has 42 pieces of candy in it. You recorded the colors of each piece as it was removed from the bag and the result is given in the follow table.

3.6. PIE CHARTS

red	brown	yellow	brown	green	yellow	yellow
brown	brown	brown	brown	red	brown	red
brown	brown	blue	brown	yellow	brown	blue
green	brown	yellow	yellow	brown	brown	brown
yellow	orange	orange	yellow	yellow	red	brown
brown	blue	red	blue	yellow	yellow	red

Create a pie chart of this data.

2. You toss a coin 20 times and record whether the obverse side (heads) or reverse side (tails) fell up. The result of these twenty tosses is shown below. In the table, H means heads and T means tails.

T	H	T	T	H
H	T	T	H	T
H	H	T	T	H
T	H	H	T	H

Display a pie graph for the tosses.

3. A pair of dice, one red and one green, is rolled in the following manner. On the first toss both dice are tossed. Then the red die is tossed while the green die keeps its value to get the second sum. Then the green die is tossed while the red die keeps its value. You continue tossing alternately the red then the green, die and generate the following table.

8	11	8	6	10	9	9
7	2	5	10	11	8	5
3	6	9	7	7	9	10
7	3	3	7	9	8	7
5	4	8	11	11	10	10

Determine a pie chart for the data.

4. The scores, in the order graded, on a recent midterm examination in Statistics, a class with 30 students, are given below.

70	80	99	98	85	89
87	79	83	38	69	70
60	69	78	40	75	56
70	51	99	69	95	86
57	53	47	50	55	81

Use the grading scale below to convert the data to nominal form and create a pie graph of the result.

Score at Least	Grade
90	A
80	B
70	C
60	D
0	F

International students may prefer the following scale

Score at Least	Grade
85	Excellent
70	Good
60	Acceptable
50	Poor
0	Fail

3.7 Broken Line Graphs

WE NOW CONSIDER two more variations on the frequency table. Both come under the heading of broken line graphs. A **broken line graph** or **piecewise linear graph** is a graph more in the sense that algebra uses the term — as the picture of a function.

Actually there are three types of broken line graphs. If you have seen the stock market news they give a broken line graph for different indices or stocks. These are part of time series analysis so will not be

3.7. BROKEN LINE GRAPHS

covered. The next type, the **frequency polygon** is a graph which indicates the shape of the point distribution of the population. The third type is an **ogive** (pronounced "oh jive") which gives the shape of the cumulative distribution for the population. It is these last two which will be covered here.

Frequency Polygon

WE FIRST LOOK AT the shape of the point distribution for the population. We assume that the data are grouped into intervals, although this is sometimes done for discrete data which are not grouped. We plot the midpoints along a horizontal number line at a height equal to the frequency of the interval — although this could also be plotted using the height of the relative frequency. Recall: The relative frequency is the frequency scaled by dividing each frequency by the total number of observations; thus, the overall shape will be identical. We then connect the points plotted with straight lines. Often the midpoint of the first interval is joined to the lower left corner and the midpoint of the rightmost interval is joined to the lower right corner to complete the polygon. Usually a scale is given on one or both sides of the graph to make it easier for people to read the changes in height. The only difference between plotting the frequencies and relative frequencies is how we label the scales on either side. We will use the frequencies here.

Example 1:
The oyster data can be grouped into as few as 4 or as many as 7 intervals. Create a frequency polygon for each of these possibilities.

Answer:

We first show the sorted data as it will be easier to derive frequency tables for each case.

5.76	6.10	6.36	6.37	6.42	6.53	6.63	6.67	6.74	6.91
6.91	6.97	7.03	7.04	7.37	7.40	7.71	7.97	8.02	8.04
8.12	8.14	8.42	8.53	8.61	8.62	9.01	9.20	9.21	9.39
9.47	9.53	9.55	9.98	10.24	10.50	10.62	10.63	10.63	10.81
10.81	10.86	11.02	11.09	11.17	11.31	11.59	11.97	12.07	13.18

We determined the intervals for each case in Example 1, §2.6, which started on page 42. We will use those intervals now.

With four groups, the intervals are [5.75, 7.61), [7.61, 9.47), [9.47, 11.33), and [11.33, 13.19]. The frequency table for this situation is

Interval	[5.75, 7.61)	[7.61, 9.47)	[9.47, 11.33)	[11.33, 13.19]
Midpoint	6.68	8.54	10.40	12.26
Frequency	16	14	16	4

Remember: It does not include the right endpoint — except the last interval — so the observation 9.47 went into the third interval, not the second. We plot the points (6.68, 16), (8.54, 14), (10.40, 16) and (12.26, 4) and join the lower right corner to the first point, consecutive points and the final point with the lower right corner with straight lines to get the following frequency polygon.

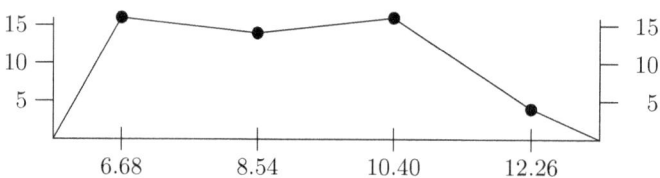

With five groups, the intervals are [5.72, 7.22), [7.22, 8.72), [8.72, 10.22), [10.22, 11.72) and [11.72, 13.22]. The frequency table becomes

Interval	[5.72, 7.22)	[7.22, 8.72)	[8.72, 10.22)	[10.22, 11.72)	[11.72, 13.22]
Midpoint	6.47	7.97	9.47	10.97	12.47
Frequency	14	12	8	13	3

3.7. BROKEN LINE GRAPHS

We plot the midpoints and interval frequencies as pairs: (6.47, 14), (7.97, 12), (9.47, 8), (10.97, 13) and (12.47, 3) and again join the lower corners and consecutive points with straight lines. We have increased the y-scale in the frequency polygon to the last to make the height of the last point easier to read.

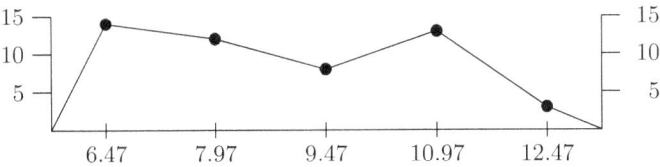

With six groups, the intervals are $[5.75, 6.99)$, $[6.99, 8.23)$, $[8.23, 9.47)$, $[9.47, 10.71)$, $[10.71, 11.95)$ and $[11.95, 13.19]$. The frequency table for these intervals is

Interval	Midpoint	Frequency
$[5.75, 6.99)$	6.37	12
$[6.99, 8.23)$	7.61	10
$[8.23, 9.47)$	8.85	8
$[9.47, 10.71)$	10.09	9
$[10.71, 11.95)$	11.33	8
$[11.95, 13.19]$	12.57	3

There were too many intervals to display them across, so we made the table vertical. Again we remind you that since the right endpoint in not included the observation 9.47 belongs in the fourth interval and not the third. We have increased the y-scale further because the highest points of the frequency polygon looked too squished and it was hard to tell the difference between them.

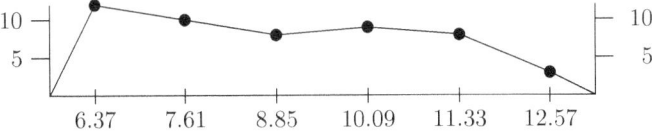

Finally, with seven groups, the intervals are [5.76, 6.82), [6.82, 7.88), [7.88, 8.94), [8.94, 10.00), [10.00, 11.06), [11.06, 12.12) and [12.12, 13.18].
These intervals make the frequency table shown below.

Interval	Midpoint	Frequency
[5.76, 6.82)	6.29	9
[6.82, 7.88)	7.35	8
[7.88, 8.94)	8.41	9
[8.94, 10.00)	9.47	13
[10.00, 11.06)	10.53	4
[11.06, 12.12)	11.59	6
[12.12, 13.18]	12.65	1

The y-scale was increased to avoid covering the tick marks.

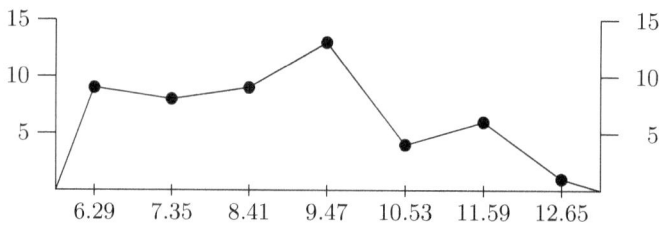

The frequency polygon above looks worse that the previous values — seven intervals seems to be too many.

Ogives

ANOTHER TYPE OF broken line graph, the **ogive** plots cumulative changes. Unlike the frequency polygons previously discussed, the points are now plotted through the endpoints of the interval. Again either frequencies or relative frequencies can be used. However, it is more common to plot relative frequencies, so the height starts at zero on the left end of the leftmost interval and rises to a height of

3.7. BROKEN LINE GRAPHS

one on the right end of the rightmost interval. Many people plot grid lines for every 20% across the graph to make it easier to read values. Then no scale is given as everyone knows the scale already. If frequencies are plotted rather than quantiles then a scale is given

Example 2:
There can be either 4 or 5 group intervals for the random gamma sample. (We leave it to you to check this statement.) Create an ogive for each of these possibilities.

Answer:
We first show the sorted data as it will be easier to derive cumulative frequencies from this.

4	5	5	5	5
8	9	9	10	12
13	13	13	14	14
15	17	17	18	22
22	24	24	27	31

The summary table below shows both the cumulative frequency and relative cumulative frequency. The midpoints of each interval are not shown as these are not needed in the ogive. Using four intervals, and making the width of each interval 7, and 5 intervals and taking the intervals as their natural width, we get the following: (Check this, as well.)

Interval	Cum Frq	Rel Cum Frq	Interval	Cum Frq	Rel Cum Frq
$[3.5, 10.5)$	9	0.36	$[4, 9.4)$	8	0.32
$[10.5, 17.5)$	18	0.72	$[9.4, 14.8)$	15	0.60
$[17.5, 24.5)$	23	0.92	$[14.8, 20.2)$	19	0.76
$[24.5, 31.5]$	25	1.00	$[20.2, 25.6)$	23	0.92
			$[25.6, 31]$	25	1.00

The ogive for four intervals is shown below. Each cumulative frequency was divided by 25, the total number of observations, to get the cumulative relative frequency.

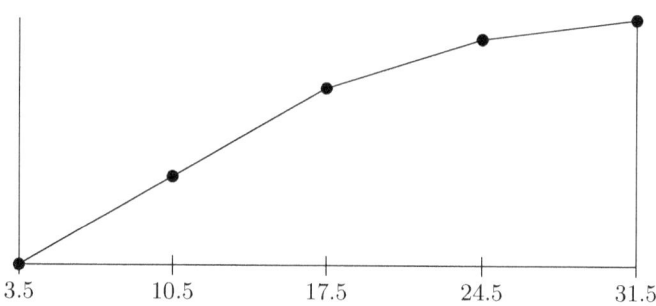

The ogive for five intervals is shown below. Again each cumulative frequency was divided by 25 to get the cumulative relative frequency.

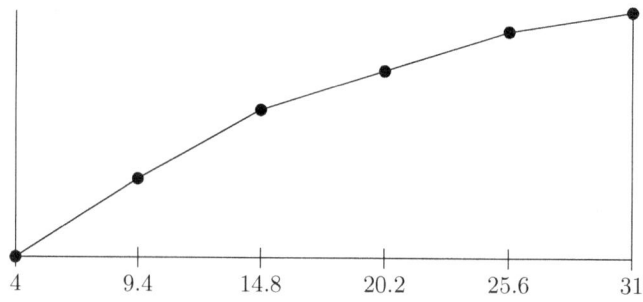

Exercises 3.7

1. One day while walking through a field you notice there seem to be a lot of stones lying on the ground. Curious about what the average size of a stone is in the field you randomly collect 30 stones. You weigh each stone and record its weight in grams. The following table shows the raw data.

3.8. HISTOGRAMS AND PARETO DIAGRAMS

110	126	75	110	126	94
90	112	111	139	131	92
102	103	96	75	67	101
104	111	100	65	92	93
106	104	93	115	109	95

Create a frequency polygon and an ogive for each possible number of intervals as calculated previously.

2. The scores, in the order graded, on a recent midterm examination in Statistics, a class with 30 students, are given below.

70	80	99	98	85	89
87	79	83	38	69	70
60	69	78	40	75	56
70	51	99	69	95	86
57	53	47	50	55	81

Create a frequency polygon and an ogive for each possible number of intervals as calculated previously.

3. Below is a sample of the monthly salary for 30 randomly selected people in mid-management in thousands of dollars.

3.08	3.18	3.13	7.41	3.73	3.05
4.17	3.58	3.36	3.27	3.36	4.74
3.32	3.61	4.26	3.02	3.45	4.06
6.73	4.72	3.13	3.15	3.70	3.59
3.12	3.61	5.03	3.53	3.20	3.32

Create a frequency polygon and an ogive for each possible number of intervals as calculated previously.

3.8 Histograms and Pareto Diagrams

THERE ARE TWO more variations of the bar chart that need to be discussed. The first is used for grouped numeric data. The second for nominal data.

Histograms

WHEN THE DATA have been grouped, a special type of bar graph can be done. A **histogram** is a bar chart for grouped numeric data. The width of each bar is the width of the underlying interval. These widths do not all need to be the same. The height of each bar is the frequency, or relative frequency of the interval. Unlike the previously discussed bar graphs, scales are placed on either side of the graph and they are not placed on or in the bars.

Example 1:
Create a histogram of the oyster data using six intervals.

Answer:
  A frequency table for these intervals was given in Example 1, §3.7, on pages 83ff. We repeat that frequency table here.

Interval	Midpoint	Frequency
$[5.75, 6.99)$	6.37	12
$[6.99, 8.23)$	7.61	10
$[8.23, 9.47)$	8.85	8
$[9.47, 10.71)$	10.09	9
$[10.71, 11.95)$	11.33	8
$[11.95, 13.19]$	12.57	3

Below is the resulting histogram.

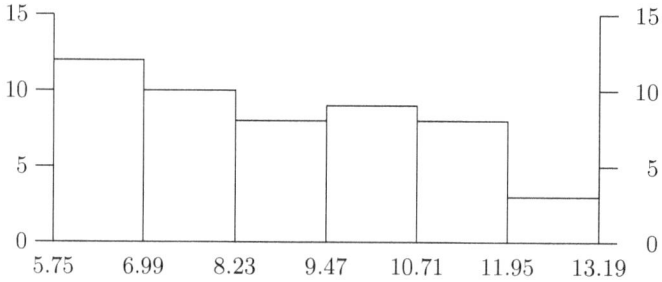

3.8. HISTOGRAMS AND PARETO DIAGRAMS

Example 2:
Make a histogram for the random gamma data using five intervals.

Answer:
The relevant portion of the frequency table of Example 2, §3.7, starting on page 87 is given below

Interval	Frq
[4, 9.4)	8
[9.4, 14.8)	7
[14.8, 20.2)	4
[20.2, 25.6)	4
[25.6, 31]	2

The histogram for this data is below.

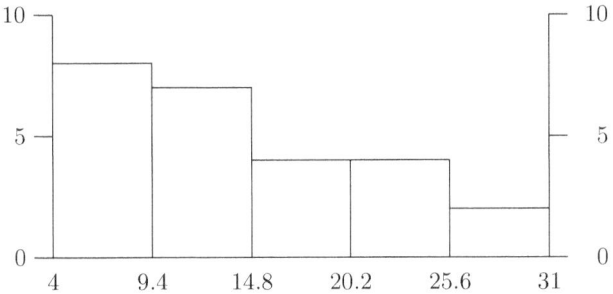

Pareto Diagram

THE LAST TYPE OF bar graph we wish to discuss is used for qualitative data, although it is sometimes used for discrete data. A **Pareto diagram** or **Pareto chart** is actually a bar chart/broken line graph combination. The categories are sorted by frequency from largest to smallest — ties can be broken arbitrarily. No space is placed between

92   CHAPTER 3. VISUALIZING DATA

the bars as an ogive will be drawn on top of the bar chart. The left scale applies to the bars. The right scale applies to the ogive.

Example 3:
Create a Pareto chart for the balloon color data.

Answer:
   A frequency table for this data was given in Example 1, §3.5, starting on page 72. This is repeated below.

blue: 6   green: 4   orange: 5   red: 5   yellow: 5

Three of the categories are tied. We will break the tie by listing them in alphabetic order. This gives the following Pareto diagram.

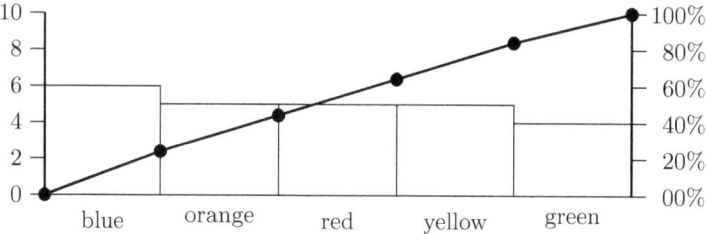

Example 4:
The results from a recent midterm in C++ for non-CS majors are shown below.

67	80	82	68	72	79	46	55	62	67	68	68
46	88	80	75	77	81	72	74	75	76	77	79
94	68	74	55	76	62	79	80	80	81	82	86
86	91	95	79			88	91	94	95		
		Unsorted						Sorted			

3.8. HISTOGRAMS AND PARETO DIAGRAMS

Create Pareto charts for the above data using the American grading scale. Compare this to the Pareto diagram using the International Grading scale. The two grading scales are given below.

	American			International	
At Least	Grade		At Least	Grade	
90	A		90	Excellent	
80	B		75	Good	
70	C		60	Acceptable	
60	D		50	Poor	
0	F		0	Fail	

Answer:

The frequency tables for the two grading scales are shown below.

Interval	Grade	Frequency		Interval	Grade	Frequency
[0, 50)	Fail	1		[0, 60)	F	2
[50, 60)	Poor	1		[60, 70)	D	4
[60, 75)	Acceptable	6		[70, 80)	C	7
[75, 90)	Good	11		[80, 90)	B	6
[90, 100]	Excellent	3		[90, 100]	A	3
	International				American	

The American Pareto chart is

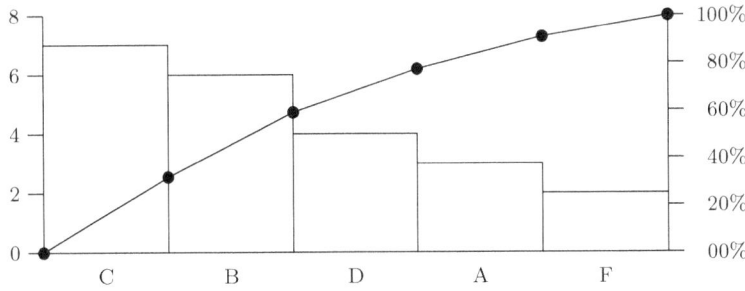

The international Pareto diagram is

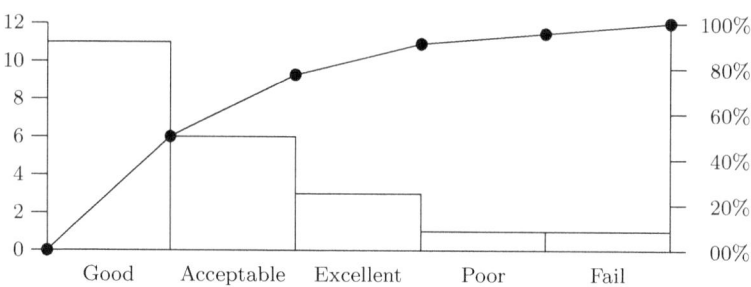

We broke the tie by placing the better grade first.

Example 5:
Create a Pareto diagram for the rune data as classified in Example 1, §3.2, which started on page 56.

Answer:

The frequency table has been repeated here for convenience.

Category	SC	MC	V	SV
Frequency	44	18	26	17

The abbreviations stood for Single Consonant, Multiple Consonant, Vowel and Semivowel. A multiple consonant has more than one sound associated with it, such as þ could be pronounced like the fricative ð in wið = with or like the allophone þ in þe = the. [Actually things were not this nice; there were about four hundred years or so when these two letters kept swapping sounds. This may be why they were eventually dropped — people couldn't keep them straight. The problem was the capitals Ð and Þ looked too similar, especially when written by hand. In fact, when introduced the spelling was wiþ and ðe.]

The Pareto diagram is given below:

3.8. HISTOGRAMS AND PARETO DIAGRAMS

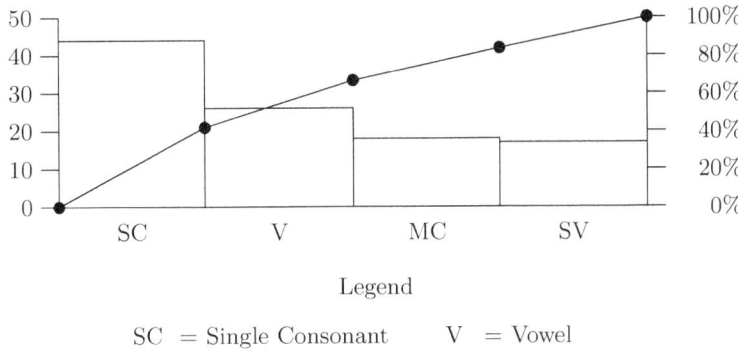

Legend

SC = Single Consonant   V = Vowel
MC = Multiple Consonant   SV = Semivowel

Exercises 3.8

1. One day while walking through a field you notice there seem to be a lot of stones lying on the ground. Curious about what the average size of a stone is in the field you randomly collect 30 stones. You weigh each stone and record its weight in grams. The following table shows the raw data.

110	126	75	110	126	94
90	112	111	139	131	92
102	103	96	75	67	101
104	111	100	65	92	93
106	104	93	115	109	95

   Create a histogram for this data using five intervals.

2. The scores, in the order graded, on a recent midterm examination in Statistics, a class with 30 students, are given below.

70	80	99	98	85	89
87	79	83	38	69	70
60	69	78	40	75	56
70	51	99	69	95	86
57	53	47	50	55	81

Create a histogram using the American grading scale. Repeat with the international grading scale. Also, using the traditional grade names, create Pareto diagrams for each case.

3. A bag of candy-coated chocolate drops has 42 pieces of candy in it. You recorded the colors of each piece as it was removed from the bag and the result is given in the follow table.

red	brown	yellow	brown	green	yellow	yellow
brown	brown	brown	brown	red	brown	red
brown	brown	blue	brown	yellow	brown	blue
green	brown	yellow	yellow	brown	brown	brown
yellow	orange	orange	yellow	yellow	red	brown
brown	blue	red	blue	yellow	yellow	red

Create a Pareto chart for these data.

4. A pair of dice, one red and one green, is rolled in the following manner. On the first toss both dice are tossed. Then the red die is tossed while the green die keeps its value to get the second sum. Then the green die is tossed while the red die keeps its value. You continue tossing alternately the red then the green, die and generate the following table.

8	11	8	6	10	9	9
7	2	5	10	11	8	5
3	6	9	7	7	9	10
7	3	3	7	9	8	7
5	4	8	11	11	10	10

Create a Pareto diagram for this data. Then create a histogram using five intervals.

3.8. HISTOGRAMS AND PARETO DIAGRAMS

5. You toss a coin 20 times and record whether the obverse side (heads) or reverse side (tails) fell up. The result of these twenty tosses is shown below. In the table, H means heads and T means tails.

T	H	T	T	H
H	T	T	H	T
H	H	T	T	H
T	H	H	T	H

Create a Pareto chart of these data.

6. The elder fuþark consisted of 24 runes, ordered as follows:

ᚠᚢᚦᚨᚱᚲᚷᚹᚺᚾᛁᛃᛇᛈᛉᛊᛏᛒᛖᛗᛚᛜᛟᛞ

You can buy sets of these made of some kind of hard plastic with a bag as they are often used in fortunetelling. They usually come with a blank tile — there is no historical evidence for this blank "rune." Suppose you bought a set. You shake the bag to mix them up, blindly reach in to get one, record the letter and return it to the bag, mixing it up again. You repeat this process 60 times and obtain the following data

The sorted order is

These can be divided into single consonants, multiple consonants, vowels and semivowels as follows:

Single Consonants	ᚱ·ᚾ·ᚲ·ᛏ·ᛒ·ᛗ·ᛕ·ᛜ·ᛞ
Multiple Consonants	ᚠ·ᚦ·ᚲ·ᚷ·ᛉ·ᛋ
Vowels	ᚢ·ᚨ·ᛁ·ᛃ·ᛖ·ᛟ
Semivowels	ᚹ·ᚺ·ᛊ

Create a Pareto diagram for this data.

Chapter 4

Quantiles

Data analysis is used to describe the key features of the population based upon the characteristics of a sample taken from that population. In this chapter we discuss rank, which is part of the centrality problem. We begin with quantiles. The chapter starts with a discussion of the extremes and then goes on to discuss the effective position of an observation. Effective position is used in both finding the percentile rank from a measurement and finding the quantile corresponding to given a percentile. Both grouped and ungrouped data are discussed. We end the chapter with a discussion of quartiles and box-and-whiskers diagrams.

4.1 Extremes

ONE IMPORTANT question that comes up is *What values are possible in the population?* This is easier to handle in terms of two questions: 1) *What is the smallest possible value?* and 2) *What is the largest possible answer?* If it is possible to conduct a census then the answers to these questions are simple. Usually, however, this is not the case. It may still be possible from the nature of the population to determine these answers, but if it isn't then we get the question *How do the sample extremes compare with the population extremes?*

If there are n observations in the sample then we use x_i to refer to the ith observation in the unsorted, raw data. Once we sort the data, $x_{(i)}$ refers to the datum which ended up in position i in the sorted data. As we have seen, it is somewhat rare for $x_i = x_{(i)}$, but it can happen.

When the data are numeric, as we are assuming in this chapter, then the sorted data are listed so that

$$x_{(1)} \leq x_{(2)} \leq \cdots \leq x_{(n)}.$$

We let $x_{\min}$ represent the **sample minimum** and $x_{\max}$ represent the **sample maximum**. From the above,

$$x_{\min} = x_{(1)} \quad \text{and} \quad x_{\max} = x_{(n)}.$$

Your instructor may insist on continuing to use $x_{(1)}$ for $x_{\min}$ and $x_{(n)}$ for $x_{\max}$. The other notations, however, do not assume that the data have been sorted. These will be the notations used throughout the rest of the book when talking about the sample minimum and maximum. The population parameter corresponding to the sample minimum is the **population minimum**, which is denoted $\boldsymbol{\mu}_{\min}$. Similarly, $\boldsymbol{\mu}_{\max}$ denotes the **population maximum**.

The sample extremes always exist. This may not be true for the population extremes. When they do exist, however, then clearly

$$\mu_{\min} \leq x_{\min} \leq x_{\max} \leq \mu_{\max}.$$

4.1. EXTREMES

When the sample is random, then

$$x_{\min} \approx \mu_{\min} \quad \text{and} \quad x_{\max} \approx \mu_{\max}.$$

Just how close the sample extremes are to the population extremes, however, depends upon several factors. Some of these are: the size of the sample; the number of possible values in the population; and the rarity of the extremes. This list is in no way exhaustive.

Example 1:
Compare the sample and population extremes for the oyster data.

Answer:
From the sorted data,

$$x_{\min} = 5.76 \text{ grams and } x_{\max} = 13.18 \text{ grams}.$$

The population extremes are completely unknown.

Example 2:
Compare the sample and population extremes for the two-success data.

Answer:
From the sorted data,

$$x_{\min} = 4 \text{ tries and } x_{\max} = 31 \text{ tries}.$$

Since there must be two successes, $\mu_{\min} = 2$ tries. It is possible that you never see two successes. Therefore, the population maximum does not exist as a definite value. People often write $\mu_{\max} < \infty$ for this.

Example 3:

A pair of dice is rolled 50 times in the normal way (not like the way done in the exercises) and the sum of the pips on the topmost faces is counted the results of these tosses is shown below.

4	7	6	5	6	7	5	10	5	7
12	4	10	3	8	9	8	5	8	7
7	9	5	8	11	8	3	8	3	10
5	6	4	6	9	4	7	9	2	8
3	7	6	10	7	10	3	6	9	6

Compare the sample and population extremes.

Answer:

The sorted data are

2	3	3	3	3	3	4	4	4	4
5	5	5	5	5	5	6	6	6	6
6	6	6	7	7	7	7	7	7	7
7	8	8	8	8	8	8	8	9	9
9	9	9	10	10	10	10	10	11	12

From the sorted data,

$$x_{\min} = 2 \text{ and } x_{\max} = 12.$$

Each die must have at least one pip and no more than six pips, so $\mu_{\min} = 2$ and $\mu_{\max} = 12$. In this case, both population extremes were observed.

Exercises 4.1

1. One day while walking through a field you notice there seem to be a lot of stones lying on the ground. Curious about what the average size of a stone is in the field you randomly collect 30 stones. You

weigh each stone and record its weight in grams. The following table shows the raw data.

110	126	75	110	126	94
90	112	111	139	131	92
102	103	96	75	67	101
104	111	100	65	92	93
106	104	93	115	109	95

Compare the sample extremes to the population extremes.

2. The scores, in the order graded, on a recent midterm examination in Statistics, a class with 30 students, are given below.

70	80	99	98	85	89
87	79	83	38	69	70
60	69	78	40	75	56
70	51	99	69	95	86
57	53	47	50	55	81

Compare the sample extremes to the population extremes.

3. Below is a sample of the monthly salary for 30 randomly selected people in mid-management in thousands of dollars.

3.08	3.18	3.13	7.41	3.73	3.05
4.17	3.58	3.36	3.27	3.36	4.74
3.32	3.61	4.26	3.02	3.45	4.06
6.73	4.72	3.13	3.15	3.70	3.59
3.12	3.61	5.03	3.53	3.20	3.32

Compare the sample extremes to the population extremes.

4. A pair of dice, one red and one green, is rolled in the following manner. On the first toss both dice are tossed. Then the red die is tossed while the green die keeps its value to get the second sum. Then the green die is tossed while the red die keeps its value. You continue tossing alternately the red then the green, die and generate the following table.

8	11	8	6	10	9	9
7	2	5	10	11	8	5
3	6	9	7	7	9	10
7	3	3	7	9	8	7
5	4	8	11	11	10	10

Compare the sample extremes to the population extremes.

4.2 Effective Position

GIVEN A SAMPLE of size n, we sort them from smallest to largest. Suppose x is any one of the observations. The question is, where along the scale from smallest possible value to largest possible value does x lie? The **effective position of x**, denoted $\#(x)$, is an estimate of this position relative to the sample.

The general problem has five cases:

1. x is an observed value in the sample

2. x lies between observed values of the sample

3. x is a population extreme

4. x lies before $x_{\min}$

5. x lies after $x_{\max}$

x is Observed

THE TRADITIONAL approach assumes no observation is repeated; that is, for all x in the sample, $\text{frq}(x) = 1$. Therefore, if $x = x_{(i)}$ then $\#(x) = i$. The problem is that every last one of our sets has

4.2. EFFECTIVE POSITION

repeated values. Most people realize that the true answer must take this frequency into account and you must average the positions to find its true effective position. Others just say that the problem cannot be solved and say it lies somewhere between the first and last occurrence in the sample. The proper method is to average the positions, which is extremely easy.

Because x is an observed value, there exists j, with $1 \leq j \leq n$, such that $x = x_{(j)}$ and if $j \neq 1$ then $x \neq x_{(j-1)}$. In other words, j is the first index where x occurs. Similarly, there exists k, with $1 \leq k \leq n$, such that $x = x_{(k)}$ and if $k \neq n$ then $x \neq x_{(k+1)}$. In other words, k is the last index where x occurs. We note that $j \leq k$. The effective position of x is thus,

$$\#(x) = \frac{j+k}{2}.$$

Note: If $j = k$ then $\text{frq}(x) = 1$, so x is unique. In this case, $\#(x) = j = k$, which agrees with the traditional method.

Example 1:
Determine the effective position for each of the following measurements taken from the random gamma data

$$5 \quad 10 \quad 13 \quad 17 \quad 31$$

Answer:
The sorted data are repeated here for convenience.

4	5	5	5	5
8	9	9	10	12
13	13	13	14	14
15	17	17	18	22
22	24	24	27	31

When $x=5$, $j=2$ and $k=5$. Therefore,
$$\#(5) = \frac{2+5}{2} = 3.5.$$

Note: Some people would only say that it lies in the interval $[2,5]$ and not give an estimate.

When $x=10$, $j=k=9$ as 10 is unique. We conclude $\#(10)=9$.
When $x=13$, $j=11$ and $k=13$. Thus,
$$\#(13) = \frac{11+13}{2} = 12.$$

Note: Some people would only say that it lies in the interval $[11,13]$ and not give an estimate.

When $x=17$, $j=17$ and $k=18$. Whence,
$$\#(17) = \frac{17+18}{2} = 17.5.$$

Note: Some people would only say that it lies in the interval $[17,18]$ and not give an estimate.

When $x=31$, $j=k=25$ as 31 is unique. We conclude $\#(31)=25$.

x is Between Observations

EVEN PEOPLE who realize that the effective position must be averaged for repeated observations have difficulty determining the effective position of values lying between observed values. Suppose x lies between $x_{(i)}$ and $x_{(i+1)}$. There are three traditional approaches.

1. The **averaging** method says
$$\#(x) = \frac{i+i+1}{2} = i+0.5.$$

4.2. EFFECTIVE POSITION

2. The **estimation** method uses linear interpolation between $x_{(i)}$ and $x_{(i+1)}$.

$$\#(x) = \frac{i \cdot [x_{(i+1)} - x] + (i+1) \cdot [x - x_{(i)}]}{x_{(i+1)} - x_{(i)}}.$$

3. The **nearest** method says: If $2x < x_{(i)} + x_{(i+1)}$ then $\#(x) = i$. If $2x > x_{(i)} + x_{(i+1)}$ then $\#(x) = i+1$. If $2x = x_{(i)} + x_{(i+1)}$ then there is inconsistency. Some average the values to get $i + 0.5$. Others take whichever of i and $i + 1$ is even.

The problem with *all* of these approaches is that they do not take into account that $x_{(i)}$ and/or $x_{(i+1)}$ may be repeated values; thus always get a value in the interval $[i, i+1]$ — which may not be appropriate.

The estimation method is closest to correct. It is the one usually employed by people who realize you must average the positions to get effective position. But if you are going to use linear interpolation then it is just as easy to do it correctly as it is to do it incorrectly. Since they realize that there is a need to average positions in effective position, they should realize that the effective positions should be used in the interpolation, not just use i and $i + 1$. Therefore the proper method determines $m_1 = \#(x_{(i)})$ and $m_2 = \#(x_{(i+1)})$ and uses these in the interpolation to get

$$\#(x) = \frac{m_1 \cdot [x_{(i+1)} - x] + m_2 \cdot [x - x_{(i)}]}{x_{(i+1)} - x_{(i)}}.$$

Example 2:
Determine the effective position for each of the following values based upon the random gamma data.

$$6 \quad 11 \quad 16 \quad 19 \quad 26$$

Compare your results with each of the traditional methods.

Answer:

The value $x = 6$ lies between $x_{(5)} = 5$ and $x_{(6)} = 8$. The effective position of $x_{(5)}$ was calculated in Example 1 above to be $m_1 = \#(x_{(5)}) = \#(5) = 3.5$. The effective position of $x_{(6)} = 8$ is $m_2 = 6$ as 8 is unique. Therefore,

$$\#(6) = \frac{3.5 \cdot (8-6) + 6 \cdot (6-5)}{8-5} = \frac{13}{3} \approx 4.33.$$

The averaging method gives

$$\#(6) = \frac{5+6}{2} = 5.5.$$

The estimation method gives

$$\#(6) = \frac{5 \cdot (8-6) + 6 \cdot (6-5)}{8-5} = \frac{16}{3} \approx 5.33.$$

The nearest method gives $\#(6) = 5$ as $2 \cdot 6 = 12 < 13 = 5 + 8$. Note that the proper effective position isn't in the interval $[5, 6]$, at all!

The value $x = 11$ lies between $x_{(9)} = 10$ and $x_{(10)} = 12$. Both of these are unique, so $m_1 = 9$ and $m_2 = 10$. We conclude

$$\#(11) = \frac{9 \cdot (12-11) + 10 \cdot (11-10)}{12-10} = \frac{19}{2} = 9.5.$$

The averaging method gives

$$\#(11) = \frac{9+10}{2} = 9.5.$$

The estimation method gives the same result as 11 is halfway between 10 and 12. The nearest method will either be the same, for the same reason

4.2. EFFECTIVE POSITION

as the estimation method, or will give the one which is even. That is, set $\#(11) = 10$, because 10 is even while 9 is odd.

The value $x = 16$ lies between $x_{(16)} = 15$ and $x_{(17)} = 17$. The observation 15 is unique, so $m_1 = 16$. The observation 17 is not, so $m_2 = (17 + 18)/2 = 17.5$. Whereby,

$$\#(16) = \frac{16 \cdot (17 - 16) + 17.5 \cdot (16 - 15)}{17 - 15} = \frac{33.5}{2} = 16.75.$$

The averaging method gives

$$\#(16) = \frac{16 + 17}{2} = 16.5.$$

The estimation method gives the same because 16 is halfway between 15 and 17. The nearest method will either be the same, for the same reason as the estimation method, or will give the one which is even. That is, set $\#(16) = 16$, because 16 is even while 17 is odd.

The value $x = 19$ lies between $x_{(19)} = 18$ and $x_{(20)} = 22$. The observation 18 is unique, so $m_1 = 19$. The observation 22 is not unique, thereby $m_2 = (20 + 21)/2 = 20.5$. Therefore,

$$\#(19) = \frac{19 \cdot (22 - 19) + 20.5 \cdot (19 - 18)}{22 - 18} = \frac{77.5}{4} = 19.375.$$

The averaging method gives

$$\#(19) = \frac{19 + 20}{2} = 19.5.$$

The estimation method gives

$$\#(19) = \frac{19 \cdot (22 - 19) + 20 \cdot (19 - 18)}{22 - 18} = \frac{77}{4} = 19.25.$$

The nearest method gives $\#(19) = 19$ because $2 \cdot 19 = 38 < 40 = 18 + 22$.

The value $x = 26$ lies between $x_{(23)} = 24$ and $x_{(24)} = 27$. The observation 24 is not unique, so $m_1 = (22+23)/2 = 22.5$. The observation 27 is unique, so $m_2 = 24$. Thence,

$$\#(26) = \frac{22.5 \cdot (27-26) + 24 \cdot (26-24)}{27-24} = \frac{70.5}{3} = 23.5.$$

The averaging method gives

$$\#(26) = \frac{23+24}{2} = 23.5.$$

The estimation method gives

$$\#(26) = \frac{23 \cdot (27-26) + 24 \cdot (26-24)}{27-24} = \frac{71}{3} \approx 23.67.$$

The nearest method gives $\#(26) = 24$ because $2 \cdot 26 = 52 > 51 = 24 + 27$.

There are even times when the nearest method will produce the correct result — with the proper amount of repetition on each side or when averaging gives the correct answer and averaging is used as part of the nearest method. More often, however, none of the traditional methods produce the correct result.

x is a Population Extreme

THE TRADITIONAL methods assume that neither $\mu_{\min}$ nor $\mu_{\max}$ is an observed value — even when they are. Traditionally, $\#(\mu_{\min}) = 0$ and $\#(\mu_{\max}) = n + 1$. By-the-by, it does not matter whether $\mu_{\min}$ and/or $\mu_{\max}$ are known or unknown — or even whether they exist in the normal sense.

4.2. EFFECTIVE POSITION

The proper method agrees with these when the appropriate extreme is not an observed value, which includes the unknown or nonexistent cases. However, if either is an observed value then that extreme is treated the same as any other observed value, the effective position is calculated not assumed. This difference will have a large impact once we begin working with percentile rank and quantiles.

Example 3:
Determine the effective position of $\mu_{\min}$ and $\mu_{\max}$ for the oyster data.

Answer:
  Surely $\mu_{\min} > 0$, but just exactly how small is anyone's guess. It is smaller than 5.76, however, as there are certainly smaller oysters, so $\#(\mu_{\min}) = 0$. Similarly, $\mu_{\max} > 13.18$ grams, but again there is no knowing exactly how large the largest oyster is. There are 50 data in the sample, so $\#(\mu_{\max}) = 50 + 1 = 51$.

Example 4:
Determine the effective positions for the dice data introduced in Example 3, § 4.1, which started on page 101.

Answer:
  Because these are dice, we know $\mu_{\min} = 2$ as each face has at least one pip and there are two dice. This value was actually observed once, hence

$$\#(\mu_{\min}) = 1.$$

Similarly, we know that $\mu_{\max} = 12$ as there are 6 pips on one of the faces for each die. This value was also observed once. There are 50 data in the sample so

$$\#(\mu_{\max}) = 50.$$

Traditionally, both values would be treated as if they were not in the sample. Therefore, $\#(\mu_{\min}) = 0$ and $\#(\mu_{\max}) = 51$ under the traditional methods.

$x < x_{\min}$

THERE IS ONE traditional method which makes no sense. It is, *let's pretend*: let's pretend this is a new $x_{\min}$ in a larger sample and set $\#(x) = 1$ in the larger sample. Forget this, it is pure drivel.

The truth is, unless you know $\mu_{\min}$, it is impossible to find an effective position. If $x_{\min} = \mu_{\min}$ then it is equally impossible as you are looking for the effective position of something which cannot occur. We must therefore assume that $\mu_{\min}$ is known and that $\mu_{\min} < x < x_{\min}$.

By convention, we set $\#(\mu_{\min}) = 0$. The method then becomes: 1) Determine $\#(x_{\min})$ and 2) Interpolate between $\mu_{\min}$ and $x_{\min}$. The fact that $\#(\mu_{\min}) = 0$ makes the formula

$$\#(x) = \frac{\#(x_{\min}) \cdot [x - \mu_{\min}]}{x_{\min} - \mu_{\min}}.$$

Example 5:
Determine the effective position of $x = 3$ from the two-successes data.

Answer:
Because there must be two successes, $\mu_{\min} = 2$. From the data (see § 3.1, Example 1, pp 50ff) we see $x_{\min} = 4$ has $\text{frq}(x_{\min}) = 5$, so $\#(x_{\min}) = (1+5)/2 = 3$. Therefore,

$$\#(3) = \frac{3 \cdot (3-2)}{4-2} = \frac{3}{2} = 1.5.$$

4.2. EFFECTIVE POSITION

The traditional methods — not the let's pretend, but the others, would assume $\text{frq}(x_{\min}) = 1$, even though it isn't. Then the averaging method gives

$$\#(3) = \frac{0+1}{2} = 0.5.$$

The estimation method agrees because 3 is halfway between 2 and 4. The nearest method would agree with this for the same reason as the estimation method. But if they did not use that then this is where the let's pretend comes into play, and I won't go there.

Example 6:
Determine the effective position for a score of 35 — the minimum allowed international grade for transcripts. Use the C++ for non-CS majors data introduced in Example 4, §3.8, which started on page 92.

Answer:
For convenience, we have repeated the sorted data below.

46	55	62	67	68	68
72	74	75	76	77	79
79	80	80	81	82	86
88	91	94	95		

It is possible to miss every question, so $\mu_{\min} = 0$. From the data, $x_{\min} = 46$ is unique, so $\#(x_{\min}) = 1$. Therefore,

$$\#(35) = \frac{1 \cdot (35 - 0)}{46 - 0} = \frac{35}{46} \approx 0.76.$$

The averaging method would give

$$\#(35) = \frac{0+1}{2} = 0.5.$$

The estimation method would agree with the proper method because $x_{\min}$ is unique. The nearest method would give $\#(35) = 1$ because $2 \cdot 35 > 0 + 46$.

$x > x_{max}$

THE SAME PEOPLE who play *let's pretend* in the small case play a similar game in this case, pretending x is a new x_{max} in a larger sample and setting $\#(x) = n+1$. Again, this is silly talk and should be forgotten.

If $\mu_{max} = x_{max}$ then it is impossible as you are looking for the effective position of something which cannot occur. There are actually two possibilities here that will produce meaningful results: 1) If μ_{max} is a known finite value and 2) if μ_{max} is larger than *any* finite value. Some people write $\mu_{max} = \infty$ for this last case, but infinity is not a number. What they *really* are trying to say is that μ_{max} has no upper limit. In both cases, the convention is to make $\#(\mu_{max}) = n+1$.

Case 1:

Suppose μ_{max} is a known finite value and $x_{max} < x < \mu_{max}$. The method then becomes: 1) Determine $\#(x_{max})$ and 2) Interpolate between x_{max} and μ_{max}. The fact that $\#(\mu_{max}) = n+1$ makes the formula

$$\#(x) = \frac{\#(x_{max}) \cdot [\mu_{max} - x] + (n+1) \cdot [x - x_{max}]}{\mu_{max} - x_{max}}.$$

Example 7:

Determine the effective position for a score of 99 based upon the C++ for non-CS majors data.

Answer:

It is possible to get every question correct, so $\mu_{max} = 100$. By convention, $\#(\mu_{max}) = n + 1 = 23$, which is already included in the formula. By the sorted data, $x_{max} = 95$ and this is unique, so $\#(x_{max}) = 22$. We

4.2. EFFECTIVE POSITION

conclude

$$\#(99) = \frac{22 \cdot (100 - 99) + 23 \cdot (99 - 95)}{100 - 95} = \frac{114}{5} = 22.8.$$

The averaging method would give

$$\#(99) = \frac{22 + 23}{2} = 22.5.$$

The estimation method would agree with the proper method because $x_{\max}$ is unique. Because 99 is more than halfway, the nearest method would play let's pretend and I won't go there.

Case 2:

Since $\mu_{\max}$ has no upper limit, $1/\mu_{\max}$ is essentially zero. We define it to be zero. There are two formulas that can be used to derive an effective position, the harmonic and the semi-harmonic methods.

With the harmonic method we interpolate between the two points $(0, 1/(n+1))$ and $(1/x_{\max}, 1/\#(x_{\max}))$ to get $1/\#(x)$. Thus,

$$\begin{aligned}
\frac{1}{\#(x)} &= \frac{\frac{1}{n+1} \cdot \left[\frac{1}{x_{\max}} - \frac{1}{x}\right] + \frac{1}{\#(x_{\max})} \cdot \frac{1}{x}}{\frac{1}{x_{\max}}} \\
&= x_{\max} \cdot \left(\frac{x - x_{\max}}{(n+1) \cdot x_{\max} \cdot x} + \frac{1}{\#(x_{\max}) \cdot x}\right) \\
&= \frac{x - x_{\max}}{(n+1) \cdot x} + \frac{x_{\max}}{\#(x_{\max}) \cdot x} \\
&= \frac{\#(x_{\max}) \cdot [x - x_{\max}] + (n+1) \cdot x_{\max}}{(n+1) \cdot \#(x_{\max}) \cdot x}.
\end{aligned}$$

Taking the reciprocal gives the result:

$$\#(x) = \frac{(n+1) \cdot \#(x_{\max}) \cdot x}{\#(x_{\max}) \cdot [x - x_{\max}] + (n+1) \cdot x_{\max}}.$$

If $x_{\max}$ is unique then this becomes

$$\#(x) = \frac{(n+1) \cdot n \cdot x}{n \cdot x + x_{\max}}.$$

With the semi-harmonic method we interpolate between the two points $(0, n+1)$ and $(1/x_{\max}, \#(x_{\max}))$ to get $\#(x)$ directly. Hence,

$$\#(x) = \frac{(n+1) \cdot \left[\dfrac{1}{x_{\max}} - \dfrac{1}{x}\right] + \dfrac{\#(x_{\max})}{x}}{\dfrac{1}{x_{\max}}}$$

$$= x_{\max} \cdot \left[\frac{(n+1) \cdot [x - x_{\max}]}{x \cdot x_{\max}} + \frac{\#(x_{\max})}{x}\right]$$

$$= \frac{(n+1) \cdot [x - x_{\max}] + \#(x_{\max}) \cdot x_{\max}}{x}.$$

If $x_{\max}$ is unique then this reduces to

$$\#(x) = \frac{(n+1) \cdot x - x_{\max}}{x}.$$

The semi-harmonic method yields slightly larger results than the harmonic method. It does not matter which you use — just pick one and be consistent in its use. If your instructor has a preference, then use that one.

4.2. EFFECTIVE POSITION

Example 8:
Determine the effective position for requiring 50 tries before seeing the second success using both methods.

Answer:
There are 100 observations and $x_{\max} = 15$ is unique, so the simplified formulas apply.

Using the harmonic method,

$$\#(x) = \frac{101 \cdot 100 \cdot 50}{100 \cdot 50 + 15} = \frac{505000}{5015} = \frac{101000}{1003} \approx 100.6979 \approx 100.70.$$

Using the semi-harmonic method,

$$\#(x) = \frac{101 \cdot 50 - 15}{50} = \frac{5035}{50} = \frac{1007}{10} = 100.7.$$

Although they rounded the same, the semi-harmonic method did produce a slightly larger result.

Exercises 4.2

1. One day while walking through a field you notice there seem to be a lot of stones lying on the ground. Curious about what the average size of a stone is in the field you randomly collect 30 stones. You weigh each stone and record its weight in grams. The following table shows the raw data.

110	126	75	110	126	94
90	112	111	139	131	92
102	103	96	75	67	101
104	111	100	65	92	93
106	104	93	115	109	95

Determine the effective position for $x = 70, 92, 113$ and 130. Explain why one cannot determine an effective position for $x = 50$ or $x = 145$.

2. The scores, in the order graded, on a recent midterm examination in Statistics, a class with 30 students, are given below.

70	80	99	98	85	89
87	79	83	38	69	70
60	69	78	40	75	56
70	51	99	69	95	86
57	53	47	50	55	81

Determine the effective position for $x = 35, 40, 63, 70$ and 99.5.

3. A pair of dice, one red and one green, is rolled in the following manner. On the first toss both dice are tossed. Then the red die is tossed while the green die keeps its value to get the second sum. Then the green die is tossed while the red die keeps its value. You continue tossing alternately the red then the green, die and generate the following table.

8	11	8	6	10	9	9
7	2	5	10	11	8	5
3	6	9	7	7	9	10
7	3	3	7	9	8	7
5	4	8	11	11	10	10

Determine effective positions for each observed value.

4.3 Percentile Rank

IN THIS SECTION we show an application of the effective position. The effective position is a relative ranking of each observation within the sample. The **percentile rank**, denoted $p(x)$, is an estimate of the percentage of the population whose values are less than or equal

4.3. PERCENTILE RANK

to that value. Hence, it does the same thing as the effective position, only now for the population. This ranking covers the interval $[0, 1]$, with
$$p(\mu_{\min}) = 0 \text{ and } p(\mu_{\max}) = 1.$$

There are three traditional method for calculating percentile rank corresponding to the three traditional methods for calculating the effective position. They all boil down to the same basic formula:
$$p(x) = \frac{\#(x)}{n+1}.$$

Like their counterparts in effective position, these methods assume that neither population extreme parameter is an observed value. In addition, when calculating percentile ranks for measurements which were not observed, they assume that no observation was repeated.

The proper method takes the possibility that $\mu_{\min}$ and $\mu_{\max}$ may have been observed values. Therefore, it ensures things lie in the interval $[0, 1]$ by adjusting the classical formula into
$$p(x) = \frac{\#(x) - \#(\mu_{\min})}{\#(\mu_{\max}) - \#(\mu_{\min})}$$

with the normal conventions that $\#(\mu_{\min}) = 0$ when $\mu_{\min} < x_{\min}$ and $\#(\mu_{\max}) = n + 1$ when $\mu_{\max} > x_{\max}$. In the case neither is observed, it reduces to the classical formula.

If the population minimum is unknown or nonexistent then there is a minimum percentile rank. The classical formula is
$$p_{\min} = \frac{1}{n+1}.$$

Because $\mu_{\max}$ may be an observation, or $\mathrm{frq}(x_{\min})$ may not be 1, the proper formula is

$$p_{\min} = \frac{\#(x_{\min})}{\#(\mu_{\max})}.$$

If the population maximum is unknown or nonexistent then there is a maximum percentile rank. The classical formula is

$$p_{\max} = \frac{n}{n+1}.$$

Because $\mu_{\min}$ may be an observation, or $\mathrm{frq}(x_{\max})$ may not be 1, the proper formula is

$$p_{\max} = \frac{\#(x_{\max}) - \#(\mu_{\min})}{n + 1 - \#(\mu_{\min})}.$$

Example 1:
Determine the minimum and maximum percentile ranks for the oyster data.

Answer:
There are 50 data. The sample extremes are $x_{\min} = 5.76$ grams and $x_{\max} = 13.18$ grams. Both of these have frequency equal to one. The population extremes do not exist because neither is well-defined. Therefore, the classical formula applies — the proper formula reduces to the classical one. We conclude

$$p_{\min} = \frac{1}{51} \approx 0.0196 = 1.96\% \text{ and } p_{\max} = \frac{50}{51} \approx 0.9804 = 98.04\%.$$

The classical formulas for $p_{\min}$ and $p_{\max}$ always sum to one. This is not true for the proper formulas as the frequencies of the sample extremes do not need to be the same and perhaps one, or both, of

4.3. PERCENTILE RANK

the population extremes is an observed value so that minimum or maximum does not apply in the classical sense.

Example 2:
Determine the percentile rank for each of the following measurements. Assume they all come from the random gamma introduced in Example 1, §3.3, which started on page 63.

$$3 \quad 5 \quad 7 \quad 10 \quad 13.5 \quad 25 \quad 30 \quad 40$$

Answer:
The sorted data are repeated here for convenience.

$$\begin{array}{ccccc} 4 & 5 & 5 & 5 & 5 \\ 8 & 9 & 9 & 10 & 12 \\ 13 & 13 & 13 & 14 & 14 \\ 15 & 17 & 17 & 18 & 22 \\ 22 & 24 & 24 & 27 & 31 \end{array}$$

The effective position of $x = 3$ is found by interpolating between $\mu_{\min}$ and $x_{\min}$. From the above,

$$\#(3) = \frac{1 \cdot (3-0)}{4-0} = \frac{3}{4} = 0.75$$

This makes the percentile rank

$$p(3) = \frac{0.75 - 0}{26 - 0} = \frac{3}{104} \approx 0.0288 = 2.88\%.$$

For all gamma distributions, $\mu_{\min} = 0$ and $\mu_{\max}$ has no upper limit.[1] Neither is an observed value; hence, $\#(\mu_{\min}) = 0$ and $\#(\mu_{\max}) = 25 + 1 = 26$. From the sorted data, $x_{\min} = 4$ is unique so $\#(x_{\min}) = 1$ and $x_{\max}$ is also unique so $\#(x_{\max}) = 25$.

---

[1] You are not expected to know this, so I am merely stating it.

By the averaging method, $\#(3) = (0+1)/2 = 0.5$. This would make the percentile rank

$$p(3) = \frac{0.5}{26} = \frac{1}{52} \approx 0.0192 = 1.92\%.$$

The estimation method would agree with the proper method because $x_{\min}$ is unique.

The nearest method would take $\#(3) = 1$ and would say one of the following

$$p(3) = \frac{1}{26} \approx 0.0385 = 3.85\% \quad \text{or} \quad p(3) = \frac{1}{27} \approx 0.0370 = 3.70\%.$$

The latter estimate comes from: let's pretend the sample is larger than it is and this is the new minimum.

The effective position of $x = 5$ is $\#(5) = (2+5)/2 = 3.5$. This makes the percentile rank

$$p(5) = \frac{3.5}{26} = \frac{7}{52} \approx 0.1346 = 13.46\%.$$

Most people would agree with that, but some would only say

$$\frac{2}{26} \leq p(5) \leq \frac{5}{26}$$

That is, that $p(5)$ lies somewhere between 7.69% and 19.23% and not give an actual estimate.

The value $x = 7$ lies between $x_{(5)} = 5$ and $x_{(6)} = 8$. Now, $\#(5) = 3.5$ from above and $\#(8) = 6$ as it is unique. Therefrom,

$$\#(7) = \frac{3.5 \cdot (8-7) + 6 \cdot (7-5)}{8-5} = \frac{15.5}{3} = \frac{31}{6} \approx 5.167.$$

This makes the percentile rank,

$$p(7) = \frac{31/6}{26} = \frac{31}{156} \approx 0.1987 = 19.87\%.$$

4.3. PERCENTILE RANK

Note: Using the rounded value 5.167 produces the same rounded value in this case. In general, however, you want to keep at least 2 extra digits and use 5.16667 to make sure it rounds correctly.

The averaging method makes $\#(7) = (5+6)/2 = 5.5$, so $p(7) = 5.5/26 \approx 0.2115 = 21.15\%$.

The estimation method calculates

$$\#(7) = \frac{5 \cdot (8-7) + 6 \cdot (7-5)}{8-5} = \frac{17}{3} = 5.667.$$

This makes $p(7) = (17/3)/26 = 17/78 \approx 0.2179 = 21.79\%$.

The nearest method makes $\#(7) = 6$ as $2 \cdot 7 > 5 + 8$, so $p(7) = p(8) = 6/26 \approx 0.2308 = 23.08\%$.

The value $x = 10$ is unique in the sample. Everyone agrees $\#(10) = 9$ and

$$p(10) = \frac{9}{26} \approx 0.3462 = 34.62\%.$$

The value $x = 13.5$ lies between $x_{(13)} = 13$ and $x_{(14)} = 14$. Now, $\#(13) = (11+13)/2 = 12$ and $\#(14) = (14+15)/2 = 14.5$. This makes

$$\#(13.5) = \frac{12 \cdot (14 - 13.5) + 14.5 \cdot (13.5 - 13)}{14 - 13} = 13.25.$$

From this,

$$p(13.5) = \frac{13.25}{26} = \frac{53}{104} \approx 0.5096 = 50.96\%.$$

The averaging method makes

$$\#(13.5) = (13+14)/2 = 13.5,$$

so $p(13.5) = 13.5/26 \approx 0.5192 = 51.92\%$.

The estimation method agrees because 13.5 is halfway between 13 and 14.

The nearest method would agree for the same reason as the estimation method, but some would make $\#(13.5) = 14$, because 14 is even, and then $p(13.5) = 14/26 \approx 0.5385 = 53.85\%$.

The value $x = 25$ lies between $x_{(23)} = 24$ and $x_{(24)} = 27$. Now, $\#(24) = (22+23)/2 = 22.5$ and $\#(27) = 24$, because this latter is unique. This makes
$$\#(25) = \frac{22.5 \cdot (27-25) + 24 \cdot (25-24)}{27-24} = \frac{69}{3} = 23;$$
so, $p(25) = 23/26 \approx 0.8846 = 88.46\%$.

The averaging method would make
$$\#(25) = \frac{23+24}{2} = 23.5; \text{ so, } p(25) = \frac{23.5}{26} = \frac{47}{52} \approx 0.9038 = 90.38\%.$$

The estimation method would have
$$\#(25) = \frac{23 \cdot (27-25) + 24 \cdot (25-24)}{27-24} = \frac{70}{3} \approx 23.33.$$
This would make
$$p(25) = \frac{70/3}{26} = \frac{35}{39} \approx 0.8974 = 89.74\%.$$

The nearest method gives $\#(25) = 23$ because $2 \cdot 25 < 24 + 27$ and would thus agree with the proper method.

The value $x = 30$ lies between $x_{(24)} = 27$ and $x_{(25)} = 31$, both of which are unique. Therefore,
$$\#(30) = \frac{24 \cdot (31-30) + 25 \cdot (30-27)}{31-27} = \frac{99}{4} = 24.75$$
and therefore
$$p(30) = \frac{24.75}{26} = \frac{99}{104} \approx 0.9519 = 95.19\%.$$

The averaging method makes $\#(30) = (24+25)/2 = 24.5$ so $p(30) = 24.5/26 \approx 0.9423 = 94.23\%$.

The estimation method agrees with the proper method because both $x = 27$ and $x = 31$ are unique.

4.3. PERCENTILE RANK

The nearest method has $\#(30) = 25$ because $2 \cdot 30 > 27 + 31$. Hence, $p(30) = p(31) = 25/26 \approx 0.9615 = 96.15\%$.

The value $x = 40$ is above $x_{\max} = 31$. Because $x_{\max}$ is unique, the simplified formulas apply. Using the harmonic method we get

$$\#(40) = \frac{26 \cdot 25 \cdot 40}{25 \cdot 40 + 31} = \frac{26000}{1031} \approx 25.2182.$$

This makes

$$p(40) = \frac{26000/1031}{26} = \frac{1000}{1031} \approx 0.9699 = 96.99\%.$$

Using the semi-harmonic method we get

$$\#(40) = \frac{26 \cdot 40 - 31}{40} = \frac{1009}{40} = 25.225.$$

Hence,

$$p(40) = \frac{1009/40}{26} = \frac{1009}{1040} \approx 0.9702 = 97.02\%.$$

The averaging method gives

$$\#(40) = \frac{25 + 26}{2} = 25.5 \text{ so } p(40) = \frac{25.5}{26} \approx 0.9808 = 98.08\%.$$

The estimation method would <u>probably</u> use the semi-harmonic method, as it is the simpler formula, but would certainly agree with one of the proper methods because $x_{\max}$ is unique.

The nearest method cannot handle this, so it plays let's pretend this is a new $x_{\max}$ in a larger sample and comes up with

$$p(40) = \frac{26}{27} \approx 0.9630 = 96.30\%.$$

Example 3:

Determine the percentile ranks for all possible dice rolls based upon the dice roll data of Example 3, § 4.1, which started on page 101.

Answer:

We repeat the sorted data below for convenience.

$$
\begin{array}{cccccccccc}
2 & 3 & 3 & 3 & 3 & 3 & 4 & 4 & 4 & 4 \\
5 & 5 & 5 & 5 & 5 & 5 & 6 & 6 & 6 & 6 \\
6 & 6 & 6 & 7 & 7 & 7 & 7 & 7 & 7 & 7 \\
7 & 8 & 8 & 8 & 8 & 8 & 8 & 8 & 9 & 9 \\
9 & 9 & 9 & 10 & 10 & 10 & 10 & 10 & 11 & 12 \\
\end{array}
$$

All possible values are present. There are 50 observations with $x_{\min} = \mu_{\min} = 2$ being unique, so $\#(\mu_{\min}) = \#(x_{\min}) = 1$. We also note that $x_{\max} = \mu_{\max} = 12$ is unique, so that $\#(\mu_{\max}) = \#(x_{\max}) = 50$. None of the traditional methods will work because at least one of the population extremes is in the sample — in this case, both are — so we will only use the proper method for this.

For $x = 2$,

$$p(2) = \frac{\#(2) - \#(\mu_{\min})}{\#(\mu_{\max}) - \#(\mu_{\min})} = \frac{1-1}{50-1} = 0 = 0\%,$$

as it should for the population minimum.

For $x = 3$, $\#(3) = (2+6)/2 = 4$. Therefore,

$$p(3) = \frac{4-1}{50-1} = \frac{3}{49} \approx 0.0612 = 6.12\%.$$

For $x = 4$, $\#(4) = (7+10)/2 = \frac{17}{2} = 8.5$. Whereby,

$$p(4) = \frac{8.5 - 1}{50 - 1} = \frac{15}{98} \approx 0.1531 = 15.31\%.$$

For $x = 5$, $\#(5) = (11+16)/2 = \frac{27}{2} = 13.5$. Thus,

$$p(5) = \frac{13.5 - 1}{50 - 1} = \frac{27}{98} \approx 0.2755 = 27.55\%.$$

4.3. PERCENTILE RANK

For $x = 6$, $\#(6) = (17 + 23)/2 = 20$. Thence,
$$p(6) = \frac{20 - 1}{50 - 1} = \frac{19}{49} = 0.3878 = 38.78\%.$$

For $x = 7$, $\#(7) = (24 + 31)/2 = \frac{55}{2} = 27.5$. Whence,
$$p(7) = \frac{27.5 - 1}{50 - 1} = \frac{53}{98} \approx 0.5408 = 54.08\%.$$

For $x = 8$, $\#(8) = (32 + 38)/2 = 35$. Hence,
$$p(8) = \frac{35 - 1}{50 - 1} = \frac{34}{49} \approx 0.6939 = 69.39\%.$$

For $x = 9$, $\#(9) = (39 + 43)/2 = 41$. Thereby,
$$p(9) = \frac{41 - 1}{50 - 1} = \frac{40}{49} \approx 0.8163 = 81.63\%.$$

For $x = 10$, $\#(10) = (44 + 48)/2 = 46$. Wherefore,
$$p(10) = \frac{46 - 1}{50 - 1} = \frac{45}{49} \approx 0.9184 = 91.84\%.$$

For $x = 11$, $\#(11) = 49$, as 11 is unique. Hereby,
$$p(11) = \frac{49 - 1}{50 - 1} = \frac{48}{49} \approx 0.9796 = 97.96\%.$$

For $x = 12$, we have already noted 12 is unique and $\#(12) = 50$. Heretofore,
$$p(12) = \frac{50 - 1}{50 - 1} = 1,$$
as it should for the population maximum.

Exercises 4.3

1. One day while walking through a field you notice there seem to be a lot of stones lying on the ground. Curious about what the average size of a stone is in the field you randomly collect 30 stones. You weigh each stone and record its weight in grams. The following table shows the raw data.

110	126	75	110	126	94
90	112	111	139	131	92
102	103	96	75	67	101
104	111	100	65	92	93
106	104	93	115	109	95

   Determine the percentile ranks for $x = 70, 92, 113$ and 130. Explain why one cannot determine an percentile rank for $x = 50$ or $x = 145$. What are the maximum and minimum percentile ranks which can be calculated for this sample?

2. The scores, in the order graded, on a recent midterm examination in Statistics, a class with 30 students, are given below.

70	80	99	98	85	89
87	79	83	38	69	70
60	69	78	40	75	56
70	51	99	69	95	86
57	53	47	50	55	81

   Determine the percentile for $x = 35, 40, 63, 70$ and 99.5.

3. A pair of dice, one red and one green, is rolled in the following manner. On the first toss both dice are tossed. Then the red die is tossed while the green die keeps its value to get the second sum. Then the green die is tossed while the red die keeps its value. You continue tossing alternately the red then the green, die and generate the following table.

4.3. PERCENTILE RANK

8	11	8	6	10	9	9
7	2	5	10	11	8	5
3	6	9	7	7	9	10
7	3	3	7	9	8	7
5	4	8	11	11	10	10

Determine percentile ranks for each observed value.

4. The following random sample of 72 observations comes from a different gamma distribution, so $\mu_{min} = 0$ and μ_{max} has no upper limit. The raw data are

2.15	1.92	1.92	2.20	3.67	4.55	3.32	2.71	4.87
2.09	1.13	1.09	2.12	4.89	3.03	3.17	4.74	4.44
4.61	2.78	2.97	2.78	2.88	3.66	2.55	4.40	3.54
2.18	3.43	3.84	4.60	1.76	2.85	2.27	1.38	1.24
4.22	1.67	4.51	3.50	6.69	5.92	6.42	0.95	4.57
3.10	3.26	2.95	2.45	1.95	2.06	3.10	2.03	4.98
0.86	2.61	1.92	2.78	1.01	2.64	3.14	0.86	3.68
2.71	3.06	4.33	4.63	2.47	2.92	1.38	2.81	5.71

Which sorted becomes

0.86	0.86	0.95	1.01	1.09	1.13	1.24	1.38	1.38
1.67	1.76	1.92	1.92	1.92	1.95	2.03	2.06	2.09
2.12	2.15	2.18	2.20	2.27	2.45	2.47	2.55	2.61
2.64	2.71	2.71	2.78	2.78	2.78	2.81	2.85	2.88
2.92	2.95	2.97	3.03	3.06	3.10	3.10	3.14	3.17
3.26	3.32	3.43	3.50	3.54	3.66	3.67	3.68	3.84
4.22	4.33	4.40	4.44	4.51	4.55	4.57	4.60	4.61
4.63	4.74	4.87	4.89	4.98	5.71	5.92	6.42	6.69

Determine the percentile ranks for each of the following values:

$$0.50 \quad 1.05 \quad 1.80 \quad 2.75 \quad 3.70 \quad 5.50 \quad 7.50$$

Compare the result obtained with proper method to that from the three traditional methods. On the last, use both the harmonic and semi-harmonic methods to determine percentile rank.

4.4 Quantiles

THE PERCENTILE RANK answered the question of relative ranking for a given measurement x of the population. Specifically, what percentage of the population lies below x? The **quantile of** p, denoted $q(p)$, reverses this question asking for the population value for which p is the percentile rank. There are four methods designed to determine the measurement x of the population. In addition, there are two methods which restricts themselves to the sample — in essence treating the sample as the entire population. We will look at restricting the quantile to the sample in the next section. For now we concentrate on mapping quantiles to the population.

All methods begin by determining the effective position corresponding to the percentile rank p. We will denote this as $\#q(p)$ to avoid double parentheses. The three traditional methods (averaging, estimation and nearest) assume that neither $\mu_{\min}$ nor $\mu_{\max}$ is an observed value and cannot be applied when even one of them is. For these we set
$$\#q(p) = p \cdot (n+1).$$
In the proper method, these are automatically taken care of so
$$\#q(p) = \#(\mu_{\min}) + p \cdot [\#(\mu_{\max}) - \#(\mu_{\min})].$$
or equivalently,
$$\#q(p) = p \cdot \#(\mu_{\max}) + (1-p) \cdot \#(\mu_{\min}).$$

$p = 0$ and $p = 1$

IN THE THREE traditional methods we cannot have $\mu_{\min}$ or $\mu_{\max}$ as observed values as this would violate the formula used for the

4.4. QUANTILES

effective position $\#q(p)$. That being said, in all methods $q(0) = \mu_{\min}$, provided that value is known, and $q(1) = \mu_{\max}$ under the same condition of being known.

If $\mu_{\min}$ is unknown or nonexistent then $q(0)$ cannot be evaluated and no value $p < p_{\min}$ can be evaluated, either. When $\mu_{\min}$ is unknown or nonexistent then the three traditional methods give

$$p_{\min} = \frac{1}{n+1}$$

while the proper method makes this

$$p_{\min} = \frac{\#(x_{\min})}{\#(\mu_{\max})}.$$

If $\mu_{\max}$ is unknown or nonexistent then $q(1)$ cannot be evaluated. When $\mu_{\max}$ is unknown or nonexistent then the three traditional methods give

$$p_{\max} = \frac{n}{n+1}$$

while the proper method makes this

$$p_{\max} = \frac{\#(x_{\max}) - \#(\mu_{\min})}{n + 1 - \#(\mu_{\min})}.$$

If $\mu_{\max}$ is known to have no finite upper limit, then the estimation method and the proper method will still be able to accept p values above $p_{\max}$ using either harmonic or semi-harmonic interpolation. The difference in the two methods being that the estimation method assumes that $\text{frq}(x_{\max}) = 1$ while the proper method does not.

Example 1:
Determine the quantiles associated with $p = 0$ and $p = 1$ on the

random gamma data. If this is not possible, what are p_{min} and p_{max}?

Answer:

All gamma distributions have $\mu_{min} = 0$, therefore $q(0) = 0$. Since μ_{min} is known, the minimum percentile does not apply. There is no finite upper limit to a gamma, so $q(1)$ is not defined. The averaging and nearest methods would have

$$p_{max} = \frac{25}{26} \approx 0.9615 = 96.15\%$$

As there are 25 observations in this data set. Probabilities above this are fine for the estimation and proper methods as they employ either harmonic or semi-harmonic interpolation, so their only limit is $p < 1$. The difference between them is that the estimation method assumes $\text{frq}(x_{max}) = 1$ and the proper method makes no such assumption.

$p < p_{min}$

THE TRADITIONAL formulas assume that $\mu_{min} < x_{min}$ and are not valid otherwise. Furthermore, this can only occur in the proper method when $\mu_{min} < x_{min}$. So we must assume this is true.

If μ_{min} is unknown or nonexistent then there is no way to obtain a meaningful result. We therefore assume μ_{min} is known. Also, when $p < p_{min}$ then $0 < \#q(p) < 1$ in the traditional methods and $0 < \#q(p) < \#(x_{min})$ in the proper method.

The averaging method assigns all $0 < p < p_{min}$ to the value halfway between μ_{min} and x_{min}. Therefore, for all these cases,

$$q(p) = \frac{\mu_{min} + x_{min}}{2}.$$

For the nearest method: if $2\#q(p) < 1$ ($0 < \#q(p) < 0.5$) then $q(p) = \mu_{min}$. If $2\#q(p) > 1$ ($0.5 < \#q(p) < 1$) then $q(p) = x_{min}$. If

4.4. QUANTILES

$2\#q(p) = 1$ ($\#q(p) = 0.5$) then

$$q(p) = \frac{\mu_{min} + x_{min}}{2}.$$

The estimation method interpolates between points $(0, \mu_{min})$ and $(1, x_{min})$; thus, assuming $\text{frq}(x_{min}) = 1$. Whereby,

$$q(p) = \mu_{min} \cdot [1 - \#q(p)] + x_{min} \cdot \#q(p).$$

The proper method also uses interpolation, only now it is interpolating between the points $(0, \mu_{min})$ and $(\#(x_{min}), x_{min})$ as it makes no assumption about $\text{frq}(x_{min})$. Thereby,

$$q(p) = \frac{\mu_{min} \cdot [\#(x_{min}) - \#q(p)] + x_{min} \cdot \#q(p)}{\#(x_{min})}.$$

Example 2:
Determine the quantile corresponding to $p = 0.025$ and $p = 0.01$ for the C++ midterm data.

Answer:
There are 22 scores in the data set. The minimum score is $x_{min} = 46$, which is unique, so $\#(x_{min}) = 1$. Neither $\mu_{min} = 0$ nor $\mu_{max} = 100$ was an observed result. Therefore, for all four methods

$$\#q(p) = p \cdot (n+1).$$

For $p = 0.025$, $\#q(0.025) = 0.025 \cdot (23) = 0.575$. The averaging method yields

$$q(0.025) = \frac{0 + 46}{2} = 23.$$

Because $2(0.575) > 1$, the nearest method determines $q(0.025) = x_{(1)} = 46$. Because $\text{frq}(x_{min}) = 1$, both the estimation and proper methods conclude

$$q(0.025) = 0 + 0.575 \cdot 46 = 26.45 \approx 26.$$

For $p = 0.01$, $\#q(0.01) = 0.01 \cdot 23 = 0.23$. The averaging method yields $q(0.01) = 23$ as it did before. Because $2(0.23) < 1$, the nearest method determines $q(0.01) = \mu_{\min} = 0$. Both the estimation and proper method get
$$q(0.01) = 0.23 \cdot 46 = 10.58 \approx 11$$
as $x_{\min}$ is unique.

$p > p_{\max}$

THE TRADITIONAL formulas assume that $\mu_{\max} > x_{\max}$ and are not valid otherwise. Furthermore, this situation can only occur in the proper method when $\mu_{\max} > x_{\max}$. So we must assume this is true.

If $\mu_{\max}$ is unknown or nonexistent then there is no way to get a meaningful result from the averaging or nearest method. If the reason it does not exist is because there is no finite upper limit, then the estimation and proper methods *can* handle this case. For now, however, we will consider the case $\mu_{\max}$ is known. When $p > p_{\max}$ then $n < \#q(p) < n+1$ in the traditional methods and $\#(x_{\max}) < \#q(p) < n+1$ in the proper method.

The averaging method will assign all $p_{\max} < p < 1$ to the value halfway between $x_{\max}$ and $\mu_{\max}$. Whereby, for all these cases,
$$q(p) = \frac{x_{\max} + \mu_{\max}}{2}.$$

The nearest method works as follows: If $2\#q(p) < 2n+1$ then $q(p) = x_{\max}$. If $2\#q(p) > 2n+1$ then $q(p) = \mu_{\max}$. If $2\#q(p) = 2n+1$ then
$$q(p) = \frac{x_{\max} + \mu_{\max}}{2}.$$

4.4. QUANTILES

The estimation method uses linear interpolation between the points $(n, x_{\max})$ and $(n+1, \mu_{\max})$, whence assuming $\operatorname{frq}(x_{\max}) = 1$. Therefore,
$$q(p) = x_{\max} \cdot [n + 1 - \#q(p)] + \mu_{\max} \cdot [\#q(p) - n].$$

The proper method also uses linear interpolation, only now it is interpolating between $(\#(x_{\max}), x_{\max})$ and $(n+1, \mu_{\max})$ as it makes no assumption about the frequency of $x_{\max}$. Hence,
$$q(p) = \frac{x_{\max} \cdot [n + 1 - \#q(p)] + \mu_{\max} \cdot [\#q(p) - \#(x_{\max})]}{n + 1 - \#(x_{\max})}.$$

Example 3:
Determine the quantile corresponding to $p = 0.975$ and $p = 0.99$ for the C++ midterm data.

Answer:
There are 22 scores in the data set. The maximum score is 95, which is unique, so $\#(x_{\max}) = 22$. Neither $\mu_{\min} = 0$ nor $\mu_{\max} = 100$ was an observed result. Therefore, for all four methods
$$\#q(p) = p \cdot (n+1).$$

For $p = 0.975$, $\#q(0.975) = 0.975 \cdot 23 = 22.425$. The averaging method estimates
$$q(0.975) = \frac{95 + 100}{2} = 97.5 \approx 98.$$
The nearest method determines that $q(0.975) = x_{\max} = 95$ because $2 \cdot (22.425) = 44.85 < 45$. The frequency of 95 is 1, so both the estimation method and the proper method calculate
$$q(0.975) = 95 \cdot (23 - 22.425) + 100 \cdot (22.425 - 22)$$
$$= 54.625 + 42.5 = 97.125 \approx 97.$$

For $p = 0.99$, $\#q(0.99) = 0.99 \cdot 23 = 22.77$. The averaging method still yields $q(0.99) \approx 98$. The nearest method now reports $q(0.99) = 100$ because $2 \cdot 22.77 = 45.54 > 45$. The estimation method and the proper method determine

$$q(0.99) = 95 \cdot (23 - 22.77) + 100 \cdot (22.77 - 22)$$
$$= 21.85 + 77 = 98.95 \approx 99.$$

When $\mu_{\max}$ is known to have no finite limit, then $1/\mu_{\max}$ is arbitrarily close to zero. We define it to be zero. This gives two possible methods for attacking this case: harmonic interpolation and semi-harmonic interpolation. We consider each of these approaches.

In harmonic interpolation, the estimation method uses interpolation between the points $(0, 1/(n+1))$ and $(1/x_{\max}, 1/n)$ to get $1/q(p)$. It assumes $\text{frq}(x_{\max}) = 1$, so $\#(x_{\max}) = n$. Whence,

$$\frac{1}{q(p)} = \frac{\frac{1}{x_{\max}} \cdot \left[\frac{1}{\#q(p)} - \frac{1}{n+1} \right]}{\frac{1}{n} - \frac{1}{n+1}}$$

$$= \frac{n \cdot (n+1)}{x_{\max}} \cdot \frac{n+1 - \#q(p)}{(n+1) \cdot \#q(p)}$$

$$= \frac{n \cdot [n+1 - \#q(p)]}{\#q(p) \cdot x_{\max}}.$$

Taking reciprocals, the estimation method gives

$$q(p) = \frac{\#q(p) \cdot x_{\max}}{n \cdot [n+1 - \#q(p)]}.$$

4.4. QUANTILES

The proper method does not make any assumption about the frequency of $x_{\max}$, so it uses $\#(x_{\max})$ in the calculation instead of n. That is, the harmonic interpolation is between the points $(0, 1/(n+1))$ and $(1/x_{\max}, 1/\#(x_{\max}))$ to derive $1/q(p)$. Therefore,

$$\frac{1}{q(p)} = \frac{\frac{1}{x_{\max}} \cdot \left[\frac{1}{\#q(p)} - \frac{1}{n+1}\right]}{\frac{1}{\#(x_{\max})} - \frac{1}{n+1}}$$

$$= \frac{\frac{1}{x_{\max}} \cdot \frac{n+1-\#q(p)}{(n+1)\cdot\#q(p)}}{\frac{n+1-\#(x_{\max})}{(n+1)\cdot\#(x_{\max})}}$$

$$= \frac{(n+1)\cdot\#(x_{\max})}{n+1-\#(x_{\max})} \cdot \frac{1}{x_{\max}} \cdot \frac{n+1-\#q(p)}{(n+1)\cdot\#q(p)}$$

$$= \frac{\#(x_{\max})\cdot[n+1-\#q(p)]}{[n+1-\#(x_{\max})]\cdot x_{\max}\cdot\#q(p)}$$

Taking reciprocals, the proper method gives

$$q(p) = \frac{x_{\max}\cdot\#q(p)\cdot[n+1-\#(x_{\max})]}{\#(x_{\max})\cdot[n+1-\#q(p)]}$$

This formula reduces to the previous formula for a unique $x_{\max}$, just as it should, as $\#(x_{\max}) = n$.

Example 4:
Use the harmonic method to determine $q(p)$ for $p = 0.975$ in the random gamma data.

Answer:
There are 25 observations and neither $\mu_{\min} = 0$ nor $\mu_{\max}$ was observed. Because $x_{\max} = 31$ is unique, the estimation method and the proper method will yield the same result.

The classic formula holds for $\#q(p)$, hence

$$\#q(0.975) = 0.975 \cdot 26 = 25.35.$$

Placing everything into the formula for $q(p)$ we get

$$q(0.975) = \frac{25.35 \cdot 31}{25 \cdot (26 - 25.35)} = \frac{1209}{25} = 48.36 \approx 48.$$

In semi-harmonic interpolation, the estimation method interpolates between the points $(0, n+1)$ and $(1/x_{\max}, n)$ to obtain $1/q(p)$. This means

$$\frac{1}{q(p)} = \frac{\frac{1}{x_{\max}} \cdot [\#q(p) - (n+1)]}{n - (n+1)} = \frac{n+1 - \#q(p)}{x_{\max}}.$$

so,

$$q(p) = \frac{x_{\max}}{n+1 - \#q(p)}.$$

The proper method does not assume $\operatorname{frq}(x_{\max}) = 1$, so it uses $\#(x_{\max})$ instead of n. This means

$$\frac{1}{q(p)} = \frac{\frac{1}{x_{\max}} \cdot [\#q(p) - (n+1)]}{\#(x_{\max}) - (n+1)} = \frac{n+1 - \#q(p)}{x_{\max} \cdot [n+1 - \#(x_{\max})]},$$

thereby,

$$q(p) = \frac{x_{\max} \cdot [n+1 - \#(x_{\max})]}{n+1 - \#q(p)}.$$

4.4. QUANTILES

Again this formula reduces to the previous one when x_{max} is unique as $\#(x_{max}) = n$.

Example 5:
Use the semi-harmonic method to determine $q(p)$ for $p = 0.975$ in the random gamma data.

Answer:
There are 25 observations and neither $\mu_{min} = 0$ nor μ_{max} was observed. Because $x_{max} = 31$ is unique, the estimation method and the proper method will yield the same result.

The classic formula holds for $\#q(p)$, hence

$$\#q(0.975) = 0.975 \cdot 26 = 25.35.$$

Placing everything into the formula for $q(p)$ we get

$$q(0.975) = \frac{31}{26 - 25.35} = \frac{620}{13} \approx 47.6923 \approx 48.$$

The semi-harmonic method has a simpler form but always gives slightly smaller values than the harmonic method. Use whichever method your instructor prefers.

$p_{min} \leq p \leq p_{max}$

THESE ARE THE values of p which are normally encountered. The traditional methods assume that neither μ_{min} nor μ_{max} is observed in the sample. Their formulas are not valid otherwise, so we assume $\mu_{min} < x_{min}$ and $\mu_{max} > x_{max}$. The formulas are equally valid under the cases where the population parameters are unknown or nonexistent under the axiom that only that which is possible can happen.

When $p_{\min} < p < p_{\max}$ then in the classical formula $1 \leq \#q(p) \leq n$ and under the proper formula $\#(x_{\min}) \leq \#q(p) \leq \#(x_{\max})$. In short, there exists m such that

$$m \leq \#q(p) < m+1 \text{ with } 1 \leq m \leq n.$$

We first consider the traditional methods. The formulas for all of the traditional methods assume $x_{(m)}$ and $x_{(m+1)}$ are different but are still used when they are not.

If $\#q(p) = m$ then the averaging, nearest and estimation methods all agree that $q(p) = x_{(m)}$. They differ only when $\#q(p)$ is not an integer.

The averaging method treats all $m < \#q(p) < m+1$ the same, returning

$$q(p) = \frac{x_{(m)} + x_{(m+1)}}{2}.$$

The nearest method calculates $2\#q(p)$. If $2\#q(p) < 2m + 1$ then $q(p) = x_{(m)}$. If $2\#q(p) > 2m + 1$ then $q(p) = x_{(m+1)}$. If $\#q(p) = 2m + 1$ then

$$q(p) = \frac{x_{(m)} + x_{(m+1)}}{2}.$$

Although, as mentioned before, some people will use $x_{(m)}$ when m is even and $x_{(m+1)}$ when m is odd for this last case.

The estimation method calculates

$$q(p) = x_{(m)} \cdot [m + 1 - \#q(p)] + x_{(m+1)} \cdot [\#q(p) - m].$$

The proper method determines $\#(x_{(m)})$. If $\#q(p) = \#(x_{(m)})$ then $q(p) = x_{(m)}$. Otherwise it calculates bounds between which interpolation can be used to obtain the value.

4.4. QUANTILES

If $\#(x_{(m)}) < \#q(p)$ then $y_1 = x_{(m)}$, $m_1 = \#(x_{(m)})$ and we define y_2 to be the next higher distinct observation with $m_2 = \#(y_2)$. If $\#(x_{(m)}) > \#q(p)$ then $y_2 = x_{(m)}$, $m_2 = \#(x_{(m)})$ and we define y_1 to be the next lower distinct observation with $m_1 = \#(y_1)$. In any event, $m_1 < \#q(p) < m_2$, so we interpolation between (y_1, m_1) and (y_2, m_2) to get $q(p)$. Therefore,

$$q(p) = \frac{y_1 \cdot [m_2 - \#q(p)] + y_2 \cdot [\#q(p) - m_1]}{m_2 - m_1}.$$

In the cases where $x_{(m)}$ and $x_{(m+1)}$ are unique, the estimation method is a true inverse of its percentile rank version. The averaging and nearest methods are only inverses for percentile rank when $\#q(p)$ is an integer and $x_{(m)}$ is unique — which combination is somewhat rare. Only the proper method is a true inverse of its percentile rank method.

Example 6:
Determine the quantiles corresponding to the 20th, 50th and 70th percentile ranks in the random gamma data.

Answer:
For convenience the sorted data is repeated below:

4	5	5	5	5
8	9	9	10	12
13	13	13	14	14
15	17	17	18	22
22	24	24	27	31

Neither $\mu_{\min}$ nor $\mu_{\max}$ is an observed value, so the classical formula for $\#q(p)$ applies.

For $p = 0.2$ we have

$$\#q(0.2) = 0.2 \cdot 26 = 5.2.$$

From the sorted data $x_{(5)} = 5$ and $x_{(6)} = 8$.

The averaging method calculates
$$q(0.2) = \frac{5+8}{2} = 6.5.$$

The nearest method sets $q(0.2) = 5$ as $2(5.2) < 11$.

The estimation method determines
$$q(0.2) = 5 \cdot (6 - 5.2) + 8 \cdot (5.2 - 5) = 4 + 1.6 = 5.6.$$

We observe $\#(5) = (2+5)/2 = 3.5 < 5.2$, so $y_1 = 5$, $m_1 = 3.5$, $y_2 = 8$ and $m_2 = 6$, because 8 is unique. The proper method calculates
$$q(0.2) = \frac{5 \cdot (6 - 5.2) + 8 \cdot (5.2 - 3.5)}{6 - 3.5} = \frac{4 + 13.6}{2.5} = \frac{176}{25} = 7.04.$$

For $p = 0.5$ we get
$$\#q(p) = 0.5 \cdot 26 = 13.$$

From the sorted data $x_{(13)} = 13$.

The averaging, nearest and estimation methods agree that $q(0.5) = 13$.

We note $\#(13) = (11+13)/2 = 12 < 13$, so $y_1 = 13$, $m_1 = 12$, $y_2 = 14$ and $m_2 = (14+15)/2 = 14.5$. The proper method concludes
$$q(0.5) = \frac{13 \cdot (14.5 - 13) + 14 \cdot (13 - 12)}{14.5 - 12} = \frac{19.5 + 14}{2.5} = \frac{67}{5} = 13.4.$$

For $p = 0.7$ we calculate
$$\#q(0.7) = 0.7 \cdot 26 = 18.2.$$

From the sorted data $x_{(18)} = 17$ and $x_{(19)} = 18$.

The averaging method concludes
$$q(0.7) = \frac{17 + 18}{2} = 17.5.$$

4.4. QUANTILES

The nearest method determines $q(0.7) = 17$ because $2(18.2) < 37$.
The estimation method estimates

$$q(0.7) = 17 \cdot (19 - 18.2) + 18 \cdot (18.2 - 18) = 13.6 + 3.6 = 17.2.$$

We notice $\#(17) = (17+18)/2 = 17.5 < 18.2$; so, $y_1 = 17$, $m_1 = 17.5$, $y_2 = 18$ and $m_2 = 19$ because 18 is unique. The proper method derives

$$q(0.7) = \frac{17 \cdot (19 - 18.2) + 18 \cdot (18.2 - 17.5)}{19 - 17.5}$$
$$= \frac{13.6 + 12.6}{1.5} = \frac{262}{15} \approx 17.4667.$$

Exercises 4.4

1. One day while walking through a field you notice there seem to be a lot of stones lying on the ground. Curious about what the average size of a stone is in the field you randomly collect 30 stones. You weigh each stone and record its weight in grams. The following table shows the raw data.

110	126	75	110	126	94
90	112	111	139	131	92
102	103	96	75	67	101
104	111	100	65	92	93
106	104	93	115	109	95

   Determine the quantiles corresponding to each percentile rank. If it is not possible, specify why and give the appropriate limit in percentile rank.

   $$p = 0.02, 0.20, 0.45, 0.85, 0.99$$

2. The scores, in the order graded, on a recent midterm examination in Statistics, a class with 30 students, are given below.

70	80	99	98	85	89
87	79	83	38	69	70
60	69	78	40	75	56
70	51	99	69	95	86
57	53	47	50	55	81

   Determine the quantiles corresponding to each percentile rank. If it is not possible, specify why and give the appropriate limit in percentile rank.

   $$p = 0.02,\ 0.20,\ 0.45,\ 0.85,\ 0.99$$

3. A pair of dice, one red and one green, is rolled in the following manner. On the first toss both dice are tossed. Then the red die is tossed while the green die keeps its value to get the second sum. Then the green die is tossed while the red die keeps its value. You continue tossing alternately the red then the green, die and generate the following table.

8	11	8	6	10	9	9
7	2	5	10	11	8	5
3	6	9	7	7	9	10
7	3	3	7	9	8	7
5	4	8	11	11	10	10

   Determine the quantiles corresponding to each percentile rank. If it is not possible, specify why and give the appropriate limit in percentile rank.

   $$p = 0.02,\ 0.20,\ 0.45,\ 0.85,\ 0.99$$

4. The following random sample of 72 observations comes from a different gamma distribution, so $\mu_{min} = 0$ and μ_{max} has no upper limit. The raw data are

4.5. THE SAMPLE AS POPULATION

2.15	1.92	1.92	2.20	3.67	4.55	3.32	2.71	4.87
2.09	1.13	1.09	2.12	4.89	3.03	3.17	4.74	4.44
4.61	2.78	2.97	2.78	2.88	3.66	2.55	4.40	3.54
2.18	3.43	3.84	4.60	1.76	2.85	2.27	1.38	1.24
4.22	1.67	4.51	3.50	6.69	5.92	6.42	0.95	4.57
3.10	3.26	2.95	2.45	1.95	2.06	3.10	2.03	4.98
0.86	2.61	1.92	2.78	1.01	2.64	3.14	0.86	3.68
2.71	3.06	4.33	4.63	2.47	2.92	1.38	2.81	5.71

Which sorted becomes

0.86	0.86	0.95	1.01	1.09	1.13	1.24	1.38	1.38
1.67	1.76	1.92	1.92	1.92	1.95	2.03	2.06	2.09
2.12	2.15	2.18	2.20	2.27	2.45	2.47	2.55	2.61
2.64	2.71	2.71	2.78	2.78	2.78	2.81	2.85	2.88
2.92	2.95	2.97	3.03	3.06	3.10	3.10	3.14	3.17
3.26	3.32	3.43	3.50	3.54	3.66	3.67	3.68	3.84
4.22	4.33	4.40	4.44	4.51	4.55	4.57	4.60	4.61
4.63	4.74	4.87	4.89	4.98	5.71	5.92	6.42	6.69

Determine the quantiles corresponding to each percentile rank. If it is not possible, specify why and give the appropriate limit in percentile rank.

$$p = 0.02, 0.20, 0.45, 0.85, 0.99$$

4.5 The Sample as Population

CALCULATING STATISTICS beyond the limits of the data is always speculative. This is why there are two methods for calculating quantiles and percentile rank when the population maximum is infinite. Even when the population extremes are known, who says that linear interpolation is best. In fact, it is just *easiest* and that is why it is commonly done. Grouped data is another situation because you lose the identity of the individual observations going beyond the data is extremely risky. In this section we consider methods which are restricted to the limits of the observations in the sample.

There is one method used for grouped data which handles effective position, percentile rank and quantiles. This method was the model from which the proper method was derived. The proper method itself is easily adapted to this restriction — just change references to the population extremes into their sample counterparts in the formulas. There is one traditional method used for ungrouped data, but it only calculates quantiles from percentile ranks and has no inverse as it is not a one-to-one procedure. We will cover these techniques in this section. Starting with ungrouped data and then going to the grouped method.

The Restricted Proper Method

THE EFFECTIVE POSITION of each observation is exactly the same as it was first presented in Section 4.2 for the regular proper method. You only need the portions where x is an observation, or x is between observations, as these are the only possibilities which survive.

The calculation of percentile changes slightly. It is now

$$p(x) = \frac{\#(x) - \#(x_{\min})}{\#(x_{\max}) - \#(x_{\min})}.$$

This equation naturally makes $p(x_{\min}) = 0$ and $p(x_{\max}) = 1$, as you would expect when treating the sample as a population.

Example 1:
Determine the percentile rank for each of the following values using the random gamma data.

4.5. THE SAMPLE AS POPULATION

$$5 \quad 6 \quad 11 \quad 13 \quad 19 \quad 22 \quad 25$$

Answer:

We display the sorted data for convenience.

$$
\begin{array}{ccccc}
4 & 5 & 5 & 5 & 5 \\
8 & 9 & 9 & 10 & 12 \\
13 & 13 & 13 & 14 & 14 \\
15 & 17 & 17 & 18 & 22 \\
22 & 24 & 24 & 27 & 31
\end{array}
$$

We observe that $\#(x_{min}) = 1$ and $\#(x_{max}) = 25$.
For $x = 5$, $\#(5) = (2+5)/2 = 3.5$. Therefore,

$$p(3.5) = \frac{3.5 - 1}{25 - 1} = \frac{5}{48} \approx 0.1041667 \approx 10.42\%.$$

The $x = 6$ lies between $x_{(5)} = 5$ and $x_{(6)} = 8$. We calculated $\#(5) = 3.5$ before. Now 8 is unique, so $\#(8) = 6$. This means

$$\#(6) = \frac{3.5 \cdot (8-6) + 6 \cdot (6-5)}{8 - 5} = \frac{7 + 6}{3} = \frac{13}{3}.$$

We conclude

$$p(6) = \frac{13/3 - 1}{24} = \frac{5}{36} \approx 0.1388889 \approx 13.89\%.$$

The value $x = 11$ lies between $x_{(9)} = 10$ and $x_{(10)} = 12$, both of which are unique. Because it is halfway between them, we conclude

$$\#(11) = \frac{9 + 10}{2} = 9.5$$

which makes

$$p(11) = \frac{9.5 - 1}{24} = \frac{17}{48} \approx 0.3541667 \approx 35.42\%.$$

For $x = 13$, $\#(13) = (11 + 13)/2 = 12$. This means

$$p(13) = \frac{12 - 1}{24} = \frac{11}{24} \approx 0.4583333 \approx 45.83\%.$$

The value $x = 19$ lies between $x_{(19)} = 18$ and $x_{(20)} = 22$. Now, 18 is unique, so $\#(18) = 19$, but $\#(22) = (20 + 21)/2 = 20.5$ as it appears twice. Whence,

$$\#(19) = \frac{19 \cdot (22 - 19) + 20.5 \cdot (19 - 18)}{22 - 18} = \frac{77.5}{4} = \frac{155}{8}.$$

We calculate

$$p(19) = \frac{155/8 - 1}{24} = \frac{49}{64} = 0.765625 \approx 76.56\%.$$

We have previously calculated $\#(22) = 20.5$; whereby,

$$p(22) = \frac{20.5 - 1}{24} = \frac{13}{16} = 0.8125 - 81.25\%.$$

The value $x = 25$ lies between $x_{(23)} = 24$ and $x_{(24)} = 27$. Now, $\#(24) = (22 + 23)/2 = 22.5$ and $\#(27) = 24$; thus,

$$\#(25) = \frac{22.5 \cdot (27 - 25) + 24 \cdot (25 - 24)}{27 - 24} = 23.$$

This makes

$$p(25) = \frac{23 - 1}{24} = \frac{11}{12} \approx 0.9166667 \approx 91.67\%.$$

Determining quantiles is equally easy. The equation for $\#q(p)$ is now

$$\#q(p) = \#(x_{\min}) + p \cdot [\#(x_{\max}) - \#(x_{\min})],$$

or equivalently,

$$\#q(p) = (1 - p) \cdot \#(x_{\min}) + p \cdot \#(x_{\max}),$$

but the details of the interpolation are exactly the same as they were for the case $p_{\min} \leq p \leq p_{\max}$.

4.5. THE SAMPLE AS POPULATION

Example 2:
Determine the quantiles corresponding to each of the following values in the random gamma data:

$$p = 0.01 \quad 0.20 \quad 0.50 \quad 0.70 \quad 0.99$$

Answer:

We observe that $\#(x_{min}) = 1$ and $\#(x_{max}) = 25$.
For $p = 0.01$.

$$\#q(0.01) = 1 + 0.01 \cdot (25 - 1) = 1.24.$$

Now $x_{(1)} = 4$ is unique while $x_{(2)} = 5$ has $\text{frq}(5) = 4$; thus, $\#(5) = (2+5)/2 = 3.5$. Therefore, $y_1 = 4$, $m_1 = 1$, $y_2 = 5$ and $m_2 = 3.5$. We conclude

$$q(0.01) = \frac{4 \cdot (3.5 - 1.24) + 5 \cdot (1.24 - 1)}{3.5 - 1} = \frac{9.04 + 1.2}{2.5} = \frac{512}{125} = 4.096.$$

For $p = 0.20$,

$$\#q(0.20) = 0.8 \cdot 1 + 0.2 \cdot 25 = 5.8.$$

We have previously calculated $\#(x_{(5)}) = 3.5$; also, $\#(x_{(6)}) = 6$; thus, we interpolate between $(5, 3.5)$ and $(8, 6)$ to get

$$q(0.20) = \frac{5 \cdot (6 - 5.8) + 8 \cdot (5.8 - 3.5)}{6 - 3.5} = \frac{1 + 18.4}{2.5} = \frac{194}{25} = 7.76.$$

For $p = 0.50$,

$$\#q(0.50) = \frac{1 + 25}{2} = 13.$$

Now $\#(x_{(13)}) = (11 + 13)/2 = 12 < 13$, so $y_1 = 13$, $m_1 = 12$, $y_2 = 14$ and $m_2 = (14 + 15)/2 = 14.5$. Hence,

$$q(0.50) = \frac{13 \cdot (14.5 - 13) + 14 \cdot (13 - 12)}{14.5 - 12} = \frac{19.5 + 14}{2.5} = \frac{67}{5} = 13.4.$$

For $p = 0.70$,
$$\#q(0.70) = 1 + 0.70 \cdot 24 = 17.8.$$

Now $\#(x_{(17)}) = (17 + 18)/2 = 17.5 < 17.8$; whence, $y_1 = 17$, $m_1 = 17.5$, $y_2 = 18$ and $m_2 = 19$. We calculate

$$q(0.70) = \frac{17 \cdot (19 - 17.8) + 18 \cdot (17.8 - 17.5)}{19 - 17.5}$$
$$= \frac{20.4 + 5.4}{1.5} = \frac{86}{5} = 17.2.$$

For $p = 0.99$,
$$\#q(0.99) = 0.01 + 0.99 \cdot 25 = 24.76.$$

Both $x_{(24)} = 27$ and $x_{(25)} = 31$ are unique, so we interpolate between $(27, 24)$ and $(31, 25)$ to get

$$q(0.99) = \frac{27 \cdot (25 - 24.76) + 31 \cdot (24.76 - 24)}{25 - 24} = 6.48 + 23.56 = 30.04.$$

The numbers may be different from when we treated the sample as only part of the population, but the techniques are exactly the same.

The Traditional Sample Method

WE ASSUME THERE are n data in the sample. The original definition of percentile was given only in terms of the sample. It is this: The $100p$ quantile is that observation such that np of the data are less than or equal to it and $n(1-p)$ of the data are greater than or equal to it. We assume that m is an integer. When $m = np$ then, assuming

4.5. THE SAMPLE AS POPULATION

there are no repeated measurements, both $x_{(m)}$ and $x_{(m+1)}$ satisfy this definition, so we set

$$q(p) = \frac{x_{(m)} + x_{(m+1)}}{2}.$$

When $m < np < m+1$, then, again under the assumption that no measurements are repeated, only $x_{(m+1)}$ satisfies this; hence,

$$q(p) = x_{(m+1)}.$$

Now, most books never mention that these definitions are only truly valid under the assumption that there are no repetitions. Thus, they just blindly say that one should do this to get that. Even without repetition, however, they are not one-to-one; hence, there is no inverse operation where you can determine the percentile rank from a measurement.

Example 3:
Determine the quantiles corresponding to each of the following values in the random gamma data:

$$p = 0.01 \quad 0.20 \quad 0.50 \quad 0.70 \quad 0.99$$

Answer:
There are 25 data.
For $p = 0.01$, $0.01 \cdot 25 = 0.25$; therefore, $q(0.01) = x_{(1)} = 4$.
For $p = 0.20$, $0.20 \cdot 25 = 5$; whence,

$$q(0.20) = (x_{(5)} + x_{(6)})/2 = (5+8)/2 = 6.5.$$

For $p = 0.50$, $0.50 \cdot 25 = 12.5$; thereby, $q(0.50) = x_{(13)} = 13$.
For $p = 0.70$, $0.70 \cdot 25 = 17.5$; whereby, $q(0.70) = x_{(18)} = 17$.
When $p = 0.99$, $0.99 \cdot 25 = 24.75$; hence, $q(0.99) = x_{(25)} = 31$.

Grouped Method

DEALING WITH grouped data is not like dealing with the actual observations. For one thing, the individual observations are blurred into just being "somewhere within an interval." Also, the width of the intervals, and their midpoints, are chosen to be easy to use in calculations. Quite often this means that the minimum value in the first interval is less than the minimum observed value and the maximum value in the last interval is larger than the largest value actually observed — unless the observation is a population extreme, for one cannot go beyond a population extreme. In fact, the only things which are *not* changed are the number of observations in each interval, and hence the total number of observations in the sample. Calculating measurements and percentile ranks, must therefore allow for fractional observations as one cannot locate the actual one anymore.

Suppose that an interval $[a, b]$ has k observations. These observations will be scattered somewhere within it. If we assume that the data are scattered uniformly within the interval then the number of observations between a and x is proportional to the fraction of the interval covered. This means we have

$$\text{offset} = \frac{k \cdot (x - a)}{b - a} \text{ observations accounted for.}$$

That is, the number of data is proportional to the ratio of the width of the interval from its starting point to the measurement and the overall width of the interval. We again use the original definition of percentile to determine exactly where each quantile is, or to determine the percentile rank from a measurement.

Let $t_0, t_1, \ldots, t_g$ be the endpoints of the g group intervals. Suppose there are n_i observations in the interval from t_{i-1} to t_i and that

4.5. THE SAMPLE AS POPULATION

x lies in this interval. Then

$$\#(x) = \sum_{k=1}^{i-1} n_k + \frac{n_i \cdot (x - t_{i-1})}{t_i - t_{i-1}}.$$

Furthermore, the percentile rank of x is

$$p(x) = \frac{\#(x)}{n}.$$

Observe: $p(t_0) = 0$ and $p(t_g) = 1$. This makes sense as $t_0 = x_{\min}$, as blurred by the intervals, and $t_g = x_{\max}$, for a similar reason.

The inverse problem is equally easy for

$$\#q(p) = p \cdot n.$$

We then subtract off the number of observations in each interval until we discover the interval in which x must lie

$$0 < \text{offset} = p \cdot n - \sum_{k=1}^{i-1} n_k \leq n_i$$

Then

$$q(p) = t_{i-1} + \frac{\text{offset} \cdot (t_i - t_{i-1})}{n_i}.$$

Example 4:
Using the two-success data determine the effective position and percentile rank for each of the following intervals.

$$4 \quad 8 \quad 12$$

Use six intervals to cover the entire set of observations.

Answer:
The sorted data has been repeated below for convenience.

2	2	2	2	2	2	2	2	2	3
3	3	3	3	3	3	3	3	3	3
3	3	3	3	3	4	4	4	4	4
4	4	4	4	4	4	4	4	4	4
5	5	5	5	5	5	5	5	5	5
5	5	5	5	5	5	5	6	6	6
6	6	6	6	6	6	7	7	7	7
8	8	8	8	9	9	9	9	9	10
10	10	10	10	10	11	11	11	11	12
12	14	14	14	14	14	15	16	17	28

We are using six intervals, so the common width is

$$w = \frac{28 - 2}{6} = 4.33.$$

We round this up to the nearest 0.1 to make $w = 4.4$. The starting point for generating endpoints should be

$$t_0 = \frac{2 + 28 - 6 \cdot 4.4}{2} = 1.8.$$

The first interval, however, must start at $t_0 = 2$ as this is the population minimum. This makes the intervals $[2, 6.2)$, $[6.2, 10.6)$, $[10.6, 15)$, $[15, 19.4)$, $[19.4, 23.8)$ and $[23.8, 28.2]$. The preliminary frequency table is

I_i	$[2, 6.2)$	$[6.2, 10.6)$	$[10.6, 15)$	$[15, 19.4)$	$[19.4, 23.8)$	$[23.8, 28.2]$
m_i	4.1	8.4	12.8	17.2	21.6	26
n_i	66	19	11	3	0	1

The value $x = 4$ lies in the first interval. The offset is

$$\text{offset} = \frac{66 \cdot (4 - 2)}{6.2 - 2} = \frac{220}{7} \approx 31.43.$$

Hence, $\#(4) = 220/7 \approx 31.429$. The percentile rank is

$$p(4) = \frac{220/7}{100} = \frac{11}{35} \approx 0.3143 = 31.43\%$$

4.5. THE SAMPLE AS POPULATION

The value $x = 8$ lies in the second interval. The offset is

$$\text{offset} = \frac{19 \cdot (8 - 6.2)}{10.6 - 6.2} = \frac{171}{22}.$$

Therefore,

$$\#(8) = 66 + \frac{171}{22} = \frac{1623}{22} \approx 73.77.$$

The percentile rank is

$$p(8) = \frac{1623/22}{100} = \frac{1623}{2200} \approx 0.7377 = 73.77\%.$$

The value $x = 12$ lies in the third interval. The offset is

$$\text{offset} = \frac{11 \cdot (12 - 10.6)}{15 - 10.6} = \frac{7}{2}.$$

This means

$$\#(12) = 66 + 19 + \frac{7}{2} = \frac{177}{2} = 88.5.$$

The percentile rank is

$$p(12) = \frac{177/2}{100} = \frac{177}{200} = 0.885 = 88.5\%.$$

The problem with this method is: What do you do when there are empty intervals? The method will make all percentile ranks the same as the end of the previous nonempty interval and will never place quantiles within the empty interval. What some people do to get around these problems is to combine each empty interval with a nonempty interval which touches it. This usually means there is a choice. Different choices will, of course, result in different solutions. Which is the best solution is anyone's guess. This is why I prefer to make one large interval rather than allow an empty interval. I tend to combine intervals from the right, but many people prefer to join the empty interval to the nonempty adjacent interval having the most observations. Still others divide up the empty interval between the adjacent nonempty intervals using the ratio of numbers of observations in each touching interval to decide where to split the empty interval — keeping in mind that the midpoint and endpoints of the modified intervals must be nice for calculations. Your instructor may have his/her own preference about this.

Example 5:
Using the two-success data determine the effective position and percentile rank for each of the following intervals.

$$16 \quad 20 \quad 24$$

Use six intervals to cover the entire set of observations. Combine the empty interval in each of the following ways:

1. *Attach the empty interval to the interval on the right.*

2. *Attach the empty interval to the interval on the left.*

3. *Divide the empty interval in the ratio of the numbers of observations on each side.*

Answer:

The three cases are

I_i	[2, 6.2)	[6.2, 10.6)	[10.6, 15)	[15, 19.4)	[19.4, 28.2]
m_i	4.1	8.4	12.8	17.2	23.8
n_i	66	19	11	3	1

4.5. THE SAMPLE AS POPULATION

I_i	$[2, 6.2)$	$[6.2, 10.6)$	$[10.6, 15)$	$[15, 23.8)$	$[23.8, 28.2]$
m_i	4.1	8.4	12.8	19.4	26
n_i	66	19	11	3	1
I_i	$[2, 6.2)$	$[6.2, 10.6)$	$[10.6, 15)$	$[15, 22.8)$	$[22.8, 28.2]$
m_i	4.1	8.4	12.8	18.9	25.5
n_i	66	19	11	3	1

Where we moved the endpoint slightly farther to the right than the 3/4 mark to make the midpoint nice. It could have easily been moved slightly left to 22.6 for the same purpose.

The value $x = 16$ always lies in the fourth interval. The offset will be slightly different in each case.

In case (1),
$$\text{offset} = \frac{3 \cdot (16 - 15)}{19.4 - 15} = \frac{15}{22}.$$

Thus,
$$\#(16) = 66 + 19 + 11 + \frac{15}{22} = \frac{2127}{22} \approx 96.68.$$

The percentile is,
$$p(16) = \frac{2127/22}{100} = \frac{2127}{2200} \approx 0.9668 = 96.68\%.$$

In case (2),
$$\text{offset} = \frac{3 \cdot (16 - 15)}{23.8 - 15} = \frac{15}{44}.$$

Whereby,
$$\#(16) = 66 + 19 + 11 + \frac{15}{44} = \frac{4239}{44} \approx 96.34.$$

The percentile is,
$$p(16) = \frac{4239/44}{100} = \frac{4239}{4400} \approx 0.9634 = 96.34\%.$$

In case (3),
$$\text{offset} = \frac{3 \cdot (16 - 15)}{22.8 - 15} = \frac{5}{13}.$$

Thereby,
$$\#(16) = 66 + 19 + 11 + \frac{5}{13} = \frac{1253}{13} \approx 96.38.$$
The percentile is,
$$p(16) = \frac{1253/13}{100} = \frac{1253}{1300} \approx 0.9638 = 96.38\%.$$

The value $x = 20$ lies in the fifth interval of case (1) and the fourth interval of the other two cases.

In case (1),
$$\text{offset} = \frac{1 \cdot (20 - 19.4)}{28.2 - 19.4} = \frac{3}{44}.$$
Thence,
$$\#(20) = 66 + 19 + 11 + 3 + \frac{3}{44} = \frac{4359}{44} \approx 99.07.$$
The percentile is
$$p(20) = \frac{4359/44}{100} = \frac{4359}{4400} \approx 0.9907 = 99.07\%.$$

In case (2),
$$\text{offset} = \frac{3 \cdot (20 - 15)}{23.8 - 15} = \frac{75}{44}.$$
Wherefore,
$$\#(20) = 66 + 19 + 11 + \frac{75}{44} = \frac{4299}{44} \approx 97.70.$$
The percentile is
$$p(20) = \frac{4299/44}{100} = \frac{4299}{4400} \approx 0.9770 = 97.70\%.$$

In case (3),
$$\text{offset} = \frac{3 \cdot (20 - 15)}{22.8 - 15} = \frac{25}{13}.$$

4.5. THE SAMPLE AS POPULATION

Heretofore,

$$\#(20) = 66 + 19 + 11 + \frac{25}{13} = \frac{1273}{13} \approx 97.92.$$

The percentile is

$$p(20) = \frac{1273/13}{100} = \frac{1273}{1300} \approx 0.9792 = 97.92\%.$$

The value $x = 24$ always lies in the fifth interval, however, the offset will be slightly different in each case.

In case (1),

$$\text{offset} = \frac{1 \cdot (24 - 19.4)}{28.2 - 19.4} = \frac{23}{44}.$$

We determine

$$\#(24) = 66 + 19 + 11 + 3 + \frac{23}{44} = \frac{4379}{44} \approx 99.52.$$

The percentile rank is

$$p(24) = \frac{4379/44}{100} = \frac{4379}{4400} \approx 0.9952 = 99.52\%.$$

In case (2),

$$\text{offset} = \frac{1 \cdot (24 - 23.8)}{28.2 - 23.8} = \frac{1}{22}.$$

We determine

$$\#(24) = 66 + 19 + 11 + 3 + \frac{1}{22} = \frac{2179}{22} \approx 99.05.$$

The percentile rank is

$$p(24) = \frac{2179/22}{100} = \frac{2179}{2200} \approx 0.9905 = 99.05\%.$$

In case (3),
$$\text{offset} = \frac{1 \cdot (24 - 22.8)}{28.2 - 22.8} = \frac{2}{9}.$$

We determine
$$\#(24) = 66 + 19 + 11 + 3 + \frac{2}{9} = \frac{893}{9} \approx 99.22.$$

The percentile rank is
$$p(24) = \frac{893/9}{100} = \frac{893}{900} \approx 0.9922 = 99.22\%.$$

Example 6:
Determine the quantile associated with each of the following percentile ranks.
$$p = 0.01 \quad 0.20 \quad 0.50 \quad 0.70 \quad 0.875 \quad 0.98 \quad 0.995$$

Use the following set of intervals and their frequencies. This is the two-success data.

I_i	$[2, 6.2)$	$[6.2, 10.6)$	$[10.6, 15)$	$[15, 22.8)$	$[22.8, 28.2]$
m_i	4.1	8.4	12.8	18.9	25.5
n_i	66	19	11	3	1

Answer:
There are 100 data.
When $p = 0.01$, $\#q(0.01) = 0.01 \cdot 100 = 1$. This is less than 66, so it is the offset into the first interval.
$$q(0.01) = 2 + \frac{1 \cdot (6.2 - 2)}{66} = \frac{227}{110} \approx 2.06.$$

When $p = 0.20$, $\#q(0.20) = 0.20 \cdot 100 = 20$. Again this is less than 66, so it is the offset into the first interval.
$$q(0.20) = 2 + \frac{20 \cdot (6.2 - 2)}{66} = \frac{36}{11} \approx 3.27.$$

4.5. THE SAMPLE AS POPULATION

When $p = 0.50$, $\#q(0.50) = 0.50 \cdot 100 = 50$. This is still in the first interval.
$$q(0.50) = 2 + \frac{50 \cdot (6.2 - 2)}{66} = \frac{57}{11} \approx 5.18.$$

When $p = 0.70$, $\#q(0.70) = 0.70 \cdot 100 = 70$. The offset into the second interval is
$$\text{offset} = 70 - 66 = 4 < 19.$$

We conclude
$$q(0.70) = 6.2 + \frac{4 \cdot (10.6 - 6.2)}{19} = \frac{677}{95} \approx 7.13.$$

When $p = 0.875$, $\#q(0.875) = 0.875 \cdot 100 = 87.5$. The offset into the third interval is
$$\text{offset} = 87.5 - 66 - 19 = \frac{5}{2} < 11.$$

This means
$$q(0.875) = 10.6 + \frac{5}{2} \cdot \frac{15 - 10.6}{11} = \frac{56}{5} = 11.6.$$

When $p = 0.98$, $\#q(0.98) = 0.98 \cdot 100 = 98$. The offset into the fourth interval is
$$\text{offset} = 98 - 66 - 19 - 11 = 2 < 3$$

Whence,
$$q(0.98) = 15 + \frac{2 \cdot (22.8 - 15)}{3} = \frac{101}{5} = 20.2.$$

When $p = 0.995$, $\#q(0.995) = 0.995 \cdot 100 = 99.5$. The offset into the fifth interval is
$$\text{offset} = 99.5 - 66 - 19 - 11 - 3 = \frac{1}{2} < 1.$$

There is no need to divide by 1, so
$$q(0.995) = 22.8 + \frac{1}{2} \cdot (28.2 - 22.8) = \frac{51}{2} = 25.5.$$

Exercises 4.5

1. One day while walking through a field you notice there seem to be a lot of stones lying on the ground. Curious about what the average size of a stone is in the field you randomly collect 30 stones. You weigh each stone and record its weight in grams. The following table shows the raw data.

110	126	75	110	126	94
90	112	111	139	131	92
102	103	96	75	67	101
104	111	100	65	92	93
106	104	93	115	109	95

   Determine the quantiles corresponding to each percentile rank — use both the traditional and proper methods.

   $$p = 0.02,\ 0.20,\ 0.45,\ 0.85,\ 0.99$$

   Group the data into five intervals and repeat the problem.

2. The scores, in the order graded, on a recent midterm examination in Statistics, a class with 30 students, are given below.

70	80	99	98	85	89
87	79	83	38	69	70
60	69	78	40	75	56
70	51	99	69	95	86
57	53	47	50	55	81

   Determine the quantiles corresponding to each percentile rank — use both the traditional and proper methods.

   $$p = 0.02,\ 0.20,\ 0.45,\ 0.85,\ 0.99$$

   Group the data into five intervals and repeat the problem.

3. A pair of dice, one red and one green, is rolled in the following manner. On the first toss both dice are tossed. Then the red die is tossed while the green die keeps its value to get the second sum. Then

4.5. THE SAMPLE AS POPULATION

the green die is tossed while the red die keeps its value. You continue tossing alternately the red then the green, die and generate the following table.

8	11	8	6	10	9	9
7	2	5	10	11	8	5
3	6	9	7	7	9	10
7	3	3	7	9	8	7
5	4	8	11	11	10	10

Determine the quantiles corresponding to each percentile rank — use both the traditional and proper methods.

$$p = 0.02, 0.20, 0.45, 0.85, 0.99$$

Group the data into five intervals and repeat the problem.

4. The following random sample of 72 observations comes from a different gamma distribution, so $\mu_{min} = 0$ and μ_{max} has no upper limit. The raw data are

2.15	1.92	1.92	2.20	3.67	4.55	3.32	2.71	4.87
2.09	1.13	1.09	2.12	4.89	3.03	3.17	4.74	4.44
4.61	2.78	2.97	2.78	2.88	3.66	2.55	4.40	3.54
2.18	3.43	3.84	4.60	1.76	2.85	2.27	1.38	1.24
4.22	1.67	4.51	3.50	6.69	5.92	6.42	0.95	4.57
3.10	3.26	2.95	2.45	1.95	2.06	3.10	2.03	4.98
0.86	2.61	1.92	2.78	1.01	2.64	3.14	0.86	3.68
2.71	3.06	4.33	4.63	2.47	2.92	1.38	2.81	5.71

Which sorted becomes

0.86	0.86	0.95	1.01	1.09	1.13	1.24	1.38	1.38
1.67	1.76	1.92	1.92	1.92	1.95	2.03	2.06	2.09
2.12	2.15	2.18	2.20	2.27	2.45	2.47	2.55	2.61
2.64	2.71	2.71	2.78	2.78	2.78	2.81	2.85	2.88
2.92	2.95	2.97	3.03	3.06	3.10	3.10	3.14	3.17
3.26	3.32	3.43	3.50	3.54	3.66	3.67	3.68	3.84
4.22	4.33	4.40	4.44	4.51	4.55	4.57	4.60	4.61
4.63	4.74	4.87	4.89	4.98	5.71	5.92	6.42	6.69

Determine the quantiles corresponding to each percentile rank — use both the traditional and proper methods.

$$p = 0.02, \ 0.20, \ 0.45, \ 0.85, \ 0.99$$

Group the data into six intervals and repeat the problem.

4.6 Box-and-Whiskers Diagrams

W<small>E HAVE ALL OF</small> the tools needed to create the last commonly used visualization of a sample. The **box-and-whiskers diagram** or **box plot** comes in two formats. Both forms require that we calculate the **quartiles**, the quantiles corresponding to the 25th (Q_1), 50th (Q_2 or M) and 75th (Q_3) percentile ranks. The **five-number summary** consists of $x_{\min}$, Q_1, M, Q_3 and $x_{\max}$ and the basic box plot is a visualization of this. The modified box-and-whiskers diagram is used to visually show the presence of data which are too extreme to be considered normal. Such observations are **outliers**. We will cover this after the basic box-and-whiskers diagram.

Quartiles

I<small>F NEITHER POPULATION</small> extreme is observed, then the quartiles can be calculated using any of the three traditional methods: averaging, nearest and estimation. One needs only keep in mind that Q_1 is the 25th percentile, M is the 50th percentile and Q_3 is the 75th percentile. Where the nearest method can be done in two ways when the effective address is halfway between integers.

Another traditional method used for quartiles is the **median-of-medians method**. The median-of-medians method calculates the

4.6. BOX-AND-WHISKERS DIAGRAMS

median using the averaging method. It then considers all data less than or equal to the median and calculates the first quartile as the median of these smaller values. Finally, it considers all data greater than or equal to the median and calculates the third quartile as the median of these larger values.

Both the regular and restricted proper methods can be used. The regular proper method uses the effective positions of the population extremes — even when they are not observed values in the sample. The restricted proper method uses the sample extremes for the limits, whether or not they are population extremes. In fact, if both population extremes are observed values in the sample then both proper methods are the same.

Finally, the traditional sample method can be used as it uses the sample extremes as the limits, just like the restricted proper method.

Thus we have six (up to eight, with variations) methods for calculating quartiles for ungrouped data.

The method choices are much more restricted when the data are grouped. There is only the grouped method.

Although the values may vary, the conclusions we obtain from the box plots rarely does — although it certainly could. Therefore, it does not really matter which method you use to calculate quartiles. Use the method recommended by your instructor. Personally, I use the restricted proper method when I know nothing about the population extremes and the regular proper method when I do. Of course, when the data are grouped then there is no choice on method, one must use the group method.

Example 1:
Determine the quartiles of the random gamma data. Then group the data, first into four and then into five intervals and calculate them

again.

Answer:

The sorted data are repeated here for convenience.

4	5	5	5	5
8	9	9	10	12
13	13	13	14	14
15	17	17	18	22
22	24	24	27	31

By the averaging method

$$\#(Q_1) = 0.25 \cdot 26 = 6.5;\ so,\ Q_1 = \frac{8+9}{2} = 8.5.$$

$$\#(M) = 0.5 \cdot 26 = 13;\ so,\ M = 13.$$

$$\#(Q_3) = 0.75 \cdot 26 = 19.5;\ so,\ Q_3 = \frac{18+22}{2} = 20.$$

The nearest method and estimation method would agree with the above as in all cases when interpolation is used we are looking for the midpoint. However, using the "even" version of the nearest method gives

$$Q_1 = x_{(6)} = 8 \text{ and } Q_3 = x_{(20)} = 22.$$

The median-of-medians method calculates M as above, so $M = 13$. To get Q_1, there are 13 data whose values are at most 13; thus,

$$\#(Q_1) = 0.5 \cdot (13 + 1) = 7;\ giving,\ Q_1 = x_{(7)} = 9.$$

There are 15 data whose values are 13 or more: $x_{(11)}$–$x_{(25)}$. The median is thus,

$$\#(Q_3) = \frac{11+25}{2} = 18;\ hence,\ Q_3 = x_{(18)} = 17.$$

The regular proper method calculates

$$\#(Q_1) = 0.25 \cdot 26 = 6.5;$$
$$\#(M) = 0.5 \cdot 26 = 13;$$
$$\#(Q_3) = 0.75 \cdot 26 = 19.5.$$

4.6. BOX-AND-WHISKERS DIAGRAMS

Now $x_{(6)}$ is unique, so $y_1 = 8$ and $m_1 = 6$, but $x_{(7)}$ is not unique; thus, $y_2 = 9$ and $m_2 = (7+8)/2 = 7.5$. Therefore,

$$Q_1 = \frac{8 \cdot (7.5 - 6.5) + 9 \cdot (6.5 - 6)}{7.5 - 6} = \frac{8 + 4.5}{1.5} = \frac{25}{3} \approx 8.33.$$

The value $x_{(13)}$ is also not unique, so $\#(13) = (11+13)/2 = 12 < 13$. We must have $y_1 = 13$, $m_1 = 12$, $y_2 = 14$ and $m_2 = (14+15)/2 = 14.5$. This makes

$$M = \frac{13 \cdot (14.5 - 13) + 14 \cdot (13 - 12)}{14.5 - 12} = \frac{19.5 + 14}{2.5} = \frac{67}{5} = 13.4.$$

Although $x_{(19)} = 18$ is unique, $x_{(20)} = 22$, is not. Thus, $y_1 = 18$, $m_1 = 19$, $y_2 = 22$ and $m_2 = (20+21)/2 = 20.5$. Hence,

$$Q_3 = \frac{18 \cdot (20.5 - 19.5) + 22 \cdot (19.5 - 19)}{20.5 - 19} = \frac{18 + 11}{1.5} = \frac{58}{3} \approx 19.33.$$

The restricted proper method calculates

$$\#(Q_1) = 1 + 0.25 \cdot 24 = 7;$$
$$\#(M) = 1 + 0.5 \cdot 24 = 13;$$
$$\#(Q_3) = 1 + 0.75 \cdot 24 = 19.$$

It is interesting that the same points will be used for interpolation of Q_1 (check this), but now

$$Q_1 = \frac{8 \cdot (7.5 - 7) + 9 \cdot (7 - 6)}{7.5 - 6} = \frac{4 + 9}{1.5} = \frac{26}{3} \approx 8.67.$$

The median is exactly the same as in the regular version, $M = 13.4$. Now, we have $Q_3 = 18$ as $x_{(19)} = 18$ is unique.

The traditional method of last section calculates

$$\#(Q_1) = 0.25 \cdot 25 = 6.25; \text{ so, } Q_1 = x_{(7)} = 9.$$
$$\#(M) = 0.5 \cdot 25 = 12.5; \text{ so, } M = x_{(13)} = 13.$$
$$\#(Q_3) = 0.75 \cdot 25 = 18.75; \text{ so, } Q_3 = x_{(19)} = 18.$$

When four intervals are used,
$$w = \frac{31-4}{4} = 6.75.$$
We will round this up to 7, so
$$t_0 = \frac{4+31-4\cdot 7}{2} = 3.5$$
This gives the following intervals and frequencies:

I_i	$[3.5, 10.5)$	$[10.5, 17.5)$	$[17.5, 24.5)$	$[24.5, 31.5]$
m_i	7	14	21	28
f_i	9	9	5	2

We calculate
$$\#(Q_1) = 0.25 \cdot 25 = 6.25;$$
$$\#(M) = 0.5 \cdot 25 = 12.5;$$
$$\#(Q_3) = 0.75 \cdot 25 = 18.75.$$
The first quartile is in the first interval, so
$$Q_1 = 3.5 + \frac{6.25 \cdot (10.5 - 3.5)}{9} = \frac{301}{36} \approx 8.36.$$
The median is in the second interval, with offset $= 12.5 - 9 = 3.5$; hence,
$$M = 10.5 + \frac{3.5 \cdot (17.5 - 10.5)}{9} = \frac{119}{9} \approx 13.22.$$
The third quartile is in the third interval with offset $= 18.75 - 18 = 0.75$; therefore,
$$Q_3 = 17.5 + \frac{0.75 \cdot (24.5 - 17.5)}{5} = \frac{371}{20} = 18.55.$$
When five intervals are used,
$$w = \frac{31-4}{5} = 5.4.$$

4.6. BOX-AND-WHISKERS DIAGRAMS

We'll keep this as it is already nice, The intervals and frequencies are now

I_i	$[4, 9.4)$	$[9.4, 14.8)$	$[14.8, 20.2)$	$[20.2, 25.6)$	$[25.6, 31]$
m_i	6.7	12.1	17.5	22.9	28.3
f_i	8	7	4	4	2

The effective positions are the same as before. The calculation details change, however.

The first quartile is still in the first interval, so

$$Q_1 = 4 + \frac{6.25 \cdot (9.4 - 4)}{8} = \frac{263}{32} = 8.21875 \approx 8.22.$$

The median is in the second interval with offset, offset $= 12.5 - 8 = 4.5$. Thus,

$$M = 9.4 + \frac{4.5 \cdot (14.8 - 9.4)}{7} = \frac{901}{70} \approx 12.87.$$

The third quartile is in the third interval with offset, offset $= 18.75 - 15 = 3.75$. Whereby,

$$Q_3 = 14.8 + \frac{3.75 \cdot (20.2 - 14.8)}{4} = \frac{1589}{80} = 19.8625 \approx 19.86.$$

In summary we have the following results for the various methods and their variations.

Method	Q_1	M	Q_3
Averaging	8.50	13.00	20.00
Nearest Averaging	8.50	13.00	20.00
Nearest or Even	8.00	13.00	22.00
Estimation	8.50	13.00	20.00
Median-of-Medians	9.00	13.00	17.00
Regular Proper	8.33	13.40	19.33
Restricted Proper	8.67	13.40	18.00
Traditional Sample	9.00	13.00	18.00
Four Intervals	8.36	13.22	18.55
Five Intervals	8.22	12.87	19.86

In the last example, the first quartile ranges from 8.00 up to 9.00. The median ranges from 12.87 up to 13.40. The third quartile ranges from 17.00 up to 22.00. Quite a difference, I'll agree, but we will see that there is very little difference in the conclusions when we get to box plots.

Basic Box Plot

AS MENTIONED BEFORE, the five-number summary consists of the sample extremes and the three quartiles. A **basic box plot** or **basic box-and-whiskers diagram** is a visualization of the five-number summary. We create a box plot in the following manner:

1. Plot the five points $x_{\min}$, Q_1, M, Q_3 and $x_{\max}$ along an invisible number line using a large tick mark. All tick marks should be the same size.

2. Join the ends of the tick mark corresponding to Q_1 with the corresponding ends of the tick mark representing Q_3. This makes the box with the tick mark at M looking like a stripe on the box.

3. Make the number line between each extreme and the nearest quartile visible. These are the whiskers.

It does not matter whether the number line is vertical or horizontal.

Example 2:
Draw the eight box plots generated in Example 1 above.
Answer:
 Except for the four-interval grouped data, the sample extremes are $x_{\min} = 4$ and $x_{\max} = 31$. With four intervals we get $x_{\min} = 3.5$ and $x_{\max} = 31.5$.
 The eight five-number summaries are

4.6. BOX-AND-WHISKERS DIAGRAMS

Method	$x_{\min}$	Q_1	M	Q_3	$x_{\max}$
Averaging	4.00	8.50	13.00	20.00	31.00
Median-of-medians	4.00	9.00	13.00	17.00	31.00
Nearest or Even	4.00	8.00	13.00	22.00	31.00
Restricted Proper	4.00	8.67	13.40	18.00	31.00
Regular Proper	4.00	8.33	13.40	19.33	31.00
Traditional Sample	4.00	9.00	13.00	18.00	31.00
Grouped 4 Intervals	3.50	8.36	13.22	18.55	31.50
Grouped 5 Intervals	4.00	8.22	12.87	19.86	31.00

Where the normal nearest method and the estimation method agreed with the averaging method and so were not included in the above table. The box plots are given below. Near is nearest or even.

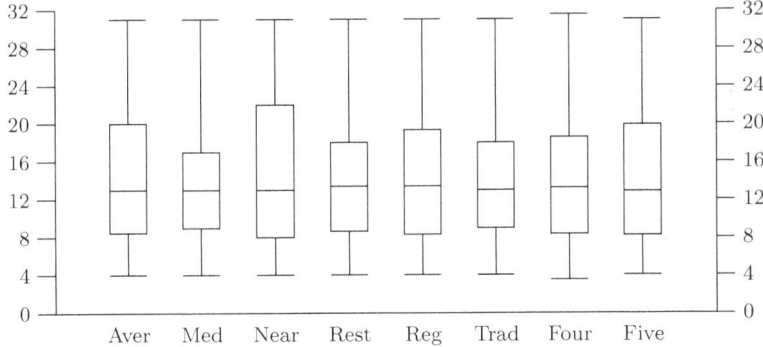

The Fence

THE LAST TIME I CHECKED, 2008, the world's smallest man was 61 cm and the world's tallest man was 270 cm — all heights to the nearest centimeter. It is possible that even a small random sample could pick one or both of these individuals. If they were both there then they come close to canceling each other out, but if only one were in the sample then that individual's presence would distort the calculation for average height well beyond what the effect should be based upon the rarity of picking that individual. It is possible, whenever measurements of any kind are taken, that a mistake can also occur which results in an observation that is far away from what would normally be expected and thus have a distorting effect on some statistic being calculated and also on the estimate for the corresponding population parameter. An **outlier** is an observation whose value is too far away from the bulk of the data to be considered normal. This begs the question: *How do we decide what values are too far away?* The answer is the fence. The **fence** is the largest interval which is classified as normal. It extends from one step below the first quartile to one step above the third quartile. So all we need to do is figure out how to calculate the size of the step.

The **range**, denoted R, is the difference between the sample extremes. Whence, $R = x_{\max} - x_{\min}$. Similarly, the **interquartile range**, denoted R_Q, is the difference between the lowest and highest quartiles. Hence, $R_Q = Q_3 - Q_1$. There are two methods for calculating the step size, one for each of these two statistics.

Most people use the **three-halves rule**. This states that

$$\text{step} = \frac{3 \cdot R_Q}{2} = 1.5 \cdot R_Q.$$

The other common method is the **three-eighths rule**. In this,

$$\text{step} = \frac{3 \cdot R}{8} = 0.375 \cdot R.$$

I suggest that the largest of these should be used for the step size. This will make the widest possible width for normal data; thus, making outliers rare as they should be.

Once the step size has been determined then we can determine the boundaries of the fence. One cannot go beyond the data, as

4.6. BOX-AND-WHISKERS DIAGRAMS

this would make no sense. Therefore, using F_L for the **lower fence boundary** and F_U for the **upper fence boundary**, we must have

$$F_L = \max(x_{\min}, Q_1 - \text{step}) \quad \text{and} \quad F_U = \min(Q_3 + \text{step}, x_{\max}).$$

The **seven-point summary** consists of the sample extremes, the boundaries of the fence, and the three quartiles: $x_{\min}$, F_L, Q_1, M, Q_3, F_U, $x_{\max}$. It is called the seven-point summary even when $F_L = x_{\min}$ and/or $F_U = x_{\max}$.

Example 3:
Determine the fence boundaries for the eight unique five-number summaries of determined in Example 1, pp 165ff. The step size should be the larger of the three-halves rule and the three-eighths rule.

Answer:
The three eighths rule, except for the four-interval grouped result, is

$$\text{step} = \frac{3 \cdot (31 - 4)}{8} = \frac{3 \cdot 27}{8} = \frac{81}{8} = 10.125.$$

For the four-interval grouped result it is

$$\text{step} = \frac{3 \cdot (31.5 - 3.5)}{8} = \frac{3 \cdot 28}{8} = \frac{21}{2} = 10.5.$$

By the averaging, normal nearest, and estimation methods, $Q_1 = 17/2$ and $Q_3 = 20$. Thus, the three-halves rule gives

$$\text{step} = \frac{3 \cdot (20 - 17/2)}{2} = \frac{69}{4} = 17.25 > 10.125.$$

We conclude

$$F_L = \max(4, 8.5 - 17.25) = 4 \text{ and } F_L = \min(31, 20 + 17.25) = 31.$$

The fence straddles all of the data, so there are no outliers.

By the median-of-medians method, $Q_1 = 9$ and $Q_3 = 17$. From the three-halves rule we obtain

$$\text{step} = \frac{3 \cdot (17 - 9)}{2} = 12 > 10.125.$$

The fence is therefore,

$$F_L = \max(4, 9 - 12) = 4 \text{ and } F_U = \min(31, 17 + 12) = 29.$$

The fence covers $[4, 29]$ and based upon the sorted data, 31 is an outlier.

By the even rounding version of the nearest method, $Q_1 = 8$ and $Q_3 = 22$. The three-halves rule makes

$$\text{step} = \frac{3 \cdot (22 - 8)}{2} = 21 > 10,125.$$

We conclude

$$F_L = \max(4, 8 - 21) = 4 \text{ and } F_U = \min(31, 22 + 21) = 31.$$

The fence straddles all of the data, so there are no outliers.

By the restricted proper method, $Q_1 = 26/3 \approx 8.67$ and $Q_3 = 18$. The three-halves rule calculates

$$\text{step} = \frac{3 \cdot (18 - 26/3)}{2} = 14 > 10.125.$$

The fence is

$$F_L = \max(4, 8.67 - 14) = 4 \text{ and } F_L = \min(31, 18 + 14) = 31.$$

The fence straddles all of the data, so there are no outliers.

The regular proper method made $Q_1 = 25/3 \approx 8.33$ and $Q_3 = 58/3 \approx 19.33$. Using the three-halves rule,

$$\text{step} = \frac{3 \cdot (58/3 - 25/3)}{2} = \frac{33}{2} = 16.5 > 10.125.$$

4.6. BOX-AND-WHISKERS DIAGRAMS

We determine the fence to be
$$F_L = \max(4, 8.33 - 16.5) = 4 \text{ and } F_U = \min(31, 19.33 + 16.5) = 31.$$
The fence straddles the entire data and there are no outliers.

The traditional sample method has $Q_1 = 9$ and $Q_3 = 18$. Placing these into the three-halves rule gives
$$\text{step} = \frac{3 \cdot (18 - 9)}{2} = \frac{27}{2} = 13.5 > 10.125.$$
Whereby the fence is
$$F_L = \max(4, 9 - 13.5) = 4 \text{ and } F_U = \min(31, 18 + 13.5) = 31.$$
The fence covers the entire sample data so there are no outliers.

When four intervals were used to group the data, $Q_1 = 301/36 \approx 8.36$ and $Q_3 = 371/20 = 18.55$. This means
$$\text{step} = \frac{3 \cdot (371/20 - 301/36)}{2} = \frac{917}{60} \approx 15.28 > 10.5$$
by the three-halves rule. Recall, the range is larger in this case. The fence is
$$F_L = \max(3.5, 8.36 - 15.28) = 3.5$$
and
$$F_U = \min(31.5, 18.55 + 15.28) = 31.5$$
There are no outliers as the fence covers even the extended values of the sample extremes.

Using five intervals makes $Q_1 = 263/32 \approx 8.22$ and $Q_3 = 1589/80 \approx 19.86$. Thereby
$$\text{step} = \frac{3 \cdot (1589/80 - 263/32)}{2} = \frac{5589}{320} \approx 17.47 > 10.125$$
using the three-halves rule. The fence is
$$F_L = \min(4, 8.22 - 17.47) = 4 \text{ and } F_U = \min(31, 19.86 + 17.47) = 31.$$
There are no outliers as the fence covers the entire data in the sample.

Only the median-of-medians method created an outlier. This will be used to create a modified box-and-whiskers diagram.

Modified Box-and-Whiskers Diagrams

WHENEVER $x_{\min} < F_L$ and/or $F_U < x_{\max}$ the sample has outliers. This does not necessarily mean that there are mistakes in the data. Rather, it indicates that there are unusual values. Values, which if kept, could distort the estimates in an exaggerated manner. A **modified box-and-whiskers diagram** or **modified box plot** is used to visually show the location of these outliers.

In a modified box plot, the whiskers stop with the fence and have no tick marks, unless this corresponds to a sample extreme, but they are labeled. If the box plot comes from ungrouped data, then each outlier is plotted with X and its value is reported. For repeated outliers a circled count is used instead of the X. In grouped data, a circled fractional count is given at the sample extreme without a tick mark. If the fractional count exceeds 1, some people prefer to place an X at the extreme and a circled count-minus-one halfway between the fence and the extreme.

Example 4:

The median-of-medians method in Example 3, pages 173ff, indicated outliers are present. Create a modified box-and-whiskers diagram for this.

Answers:
    The seven-number summary for the median-of-medians method on the random gamma data is

$$x_{\min} = 4; F_L = 4; Q_1 = 9; M = 13; Q_3 = 17; F_U = 29; x_{\max} = 31.$$

4.6. BOX-AND-WHISKERS DIAGRAMS

From the sorted data given in Example 1, starting on page 165, we see that only $x_{\max}$ is an outlier. This gives the following modified box plot.

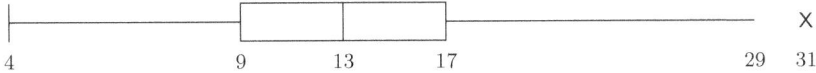

I prefer to do modified box-and-whiskers diagrams horizontally so that they can take the entire line, if needed. Many people prefer vertical box plots as was done in Example 1, starting on page 165.

Example 5:
Using the two-success data, with intervals as determined in Case (3) of Example 5, §4.5, starting on page 156. Determine if there are any outliers and draw the, possibly modified, box plot for the data.

Answers:
The frequency table for this case is repeated here for convenience

I_i	$[2, 6.2)$	$[6.2, 10.6)$	$[10.6, 15)$	$[15, 22.8)$	$[22.8, 28.2]$
m_i	4.1	8.4	12.8	18.9	25.5
n_i	66	19	11	3	1

The first quartile has $\#(Q) = 0.25 \cdot 100 = 25$ and therefore lies in the first interval.
$$Q_1 = 2 + \frac{25 \cdot (6.2 - 2)}{66} = \frac{79}{22} \approx 3.59.$$
The median has $\#(M) = 0.5 \cdot 100 = 50$ and again lies in the first interval.
$$M = 2 + \frac{50 \cdot (6.2 - 2)}{66} = \frac{57}{11} \approx 5.18.$$
The third quartile has $\#(Q_3) = 0.75 \cdot 100 = 75$, from which we see that Q_3 lies in the second interval. We determine
$$\text{offset} = 75 - 66 = 9$$

and therefore
$$Q_3 = 6.2 + \frac{9 \cdot (10.6 - 6.2)}{19} = \frac{787}{95} \approx 8.28.$$
By the three-eighths rule
$$\text{step} = \frac{3 \cdot (28.2 - 2)}{8} = \frac{393}{40} = 9.825 \approx 9.82.$$
By the three-halves rule
$$\text{step} = \frac{3 \cdot (787/95 - 79/22)}{2} = \frac{29427}{4180} \approx 7.04.$$

We conclude that the step size from the three-eighths rule should be used to calculate the fence. This means
$$F_L = \max(2, 3.59 - 9.82) = 2$$
and
$$F_U = \min(28.2, 8.28 + 9.82) = 18.11.$$
There is rounding error because both 8.28 and 9.82 were rounded down. I have used the correct value for F_U when rounded to two decimal places.

This means that the entire fifth interval, and part of the fourth interval are outliers. There is one outlier in the fifth interval, which we will represent as an X for the extreme. There are 3 observations in the fourth interval; the number of outliers is
$$\text{outliers} \approx \frac{3 \cdot (22.8 - 18.11)}{22.8 - 15} \approx 1.8.$$

We will place an outlier on the end of the fourth interval and a fractional outlier halfway between the fence and this outlier. This gives the following modified box-and-whiskers diagram.

4.6. BOX-AND-WHISKERS DIAGRAMS

We have chosen to mark the median above as it would look too crowded with the location of Q_1 being where it is. Many people alternate labels below and above. We tend to place them all below unless two consecutive labels are too close.

Exercises 4.6

1. One day while walking through a field you notice there seem to be a lot of stones lying on the ground. Curious about what the average size of a stone is in the field you randomly collect 30 stones. You weigh each stone and record its weight in grams. The following table shows the raw data.

110	126	75	110	126	94
90	112	111	139	131	92
102	103	96	75	67	101
104	111	100	65	92	93
106	104	93	115	109	95

   Create a box-and-whiskers diagram for the data. Determine the fence and, if appropriate, make a modified box plot.

2. The scores, in the order graded, on a recent midterm examination in Statistics, a class with 30 students, are given below.

70	80	99	98	85	89
87	79	83	38	69	70
60	69	78	40	75	56
70	51	99	69	95	86
57	53	47	50	55	81

   Create a box-and-whiskers diagram for the data. Determine the fence and, if appropriate, make a modified box plot.

3. Below is a sample of the monthly salary for 30 randomly selected people in mid-management in thousands of dollars.

3.08	3.18	3.13	7.41	3.73	3.05
4.17	3.58	3.36	3.27	3.36	4.74
3.32	3.61	4.26	3.02	3.45	4.06
6.73	4.72	3.13	3.15	3.70	3.59
3.12	3.61	5.03	3.53	3.20	3.32

Create a box-and-whiskers diagram for the data. Determine the fence and, if appropriate, make a modified box plot.

4. A pair of dice, one red and one green, is rolled in the following manner. On the first toss both dice are tossed. Then the red die is tossed while the green die keeps its value to get the second sum. Then the green die is tossed while the red die keeps its value. You continue tossing alternately the red then the green, die and generate the following table.

8	11	8	6	10	9	9
7	2	5	10	11	8	5
3	6	9	7	7	9	10
7	3	3	7	9	8	7
5	4	8	11	11	10	10

Create a box-and-whiskers diagram for the data. Determine the fence and, if appropriate, make a modified box plot.

5. The following random sample of 72 observations comes from a gamma distribution, so $\mu_{min} = 0$ and μ_{max} has no upper limit. The raw data are

2.15	1.92	1.92	2.20	3.67	4.55	3.32	2.71	4.87
2.09	1.13	1.09	2.12	4.89	3.03	3.17	4.74	4.44
4.61	2.78	2.97	2.78	2.88	3.66	2.55	4.40	3.54
2.18	3.43	3.84	4.60	1.76	2.85	2.27	1.38	1.24
4.22	1.67	4.51	3.50	6.69	5.92	6.42	0.95	4.57
3.10	3.26	2.95	2.45	1.95	2.06	3.10	2.03	4.98
0.86	2.61	1.92	2.78	1.01	2.64	3.14	0.86	3.68
2.71	3.06	4.33	4.63	2.47	2.92	1.38	2.81	5.71

Create a box-and-whiskers diagram for the data. Determine the fence and, if appropriate, make a modified box plot.

Chapter 5

Centrality

Centrality is the second part of location. It is concerned with determining the most central, most common, or most representative values of the population by looking at the corresponding problems in the sample. We look at each of these cases in turn. Before completing the most representative, however, we will consider the various ways to determine an average.

5.1 Center of Observations

When the data are discrete, the population can be viewed as a set of possible values. Being numeric, this set can be ordered from smallest to largest to give the **set of possible responses** (or **set of possibilities**), denoted Ω. If $|\Omega| = N$ then we can write

$$\Omega = \{\omega_1, \omega_2, \ldots, \omega_N\}.$$

The **center of possibilities**, denoted $\mathfrak{C}$ is the set consisting of the center-most values. When N is odd, there exists an integer k such that $N = 2k - 1$. Then
$$\mathfrak{C} = \{\omega_k\}.$$
When N is even, there exists k such that $N = 2k$. Then
$$\mathfrak{C} = \{\omega_k, \omega_{k+1}\}.$$

Example 1:
You have a spinner like the one shown on the right and spin the spinner counterclockwise, recording the number in the region it stops, or the previous region if it stops on a line. What are the set of possibilities and center of possibilities for this experiment?

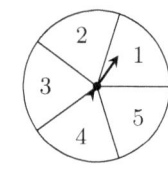

Answer:
The set of possibilities is
$$\Omega = \{1, 2, 3, 4, 5\}$$
and the center of possibilities is
$$\mathfrak{C} = \{3\}.$$

Example 2:
You toss a a pair of tetrahedral dice (four-sided dice, each die resembling a triangular pyramid) and record the sum of the pips on the bottommost faces. What are the set of possibilities and center of possibilities for this experiment?

Answer:

5.1. CENTER OF OBSERVATIONS

The set of possibilities is

$$\Omega = \{1, 2, 3, 4, 5, 6, 7, 8\}$$

and the center of possibilities is

$$\mathcal{C} = \{4, 5\}.$$

Every sample is finite and so has only a finite number of distinct observations. The **set of distinct observations**, denoted S, is assumed to be ordered. Even when the data are continuous there are only a finite number of distinct values seen in the sample, so S always exists. If there are k distinct values in S, then this ordered set is

$$S = \{y_1, y_2, \ldots, y_k\}.$$

The **center of observations**, denoted $\mathcal{C}$, is the set of center-most values in the set of distinct observations. Thus, for $k = 2m - 1$, $\mathcal{C} = \{y_m\} = y_m$ and for $k = 2m$, $\mathcal{C} = \{y_m, y_{m+1}\}$.

Example 3:
Determine the center of observations for the dice roll data introduced in Example 3, §4.1, which started on page 101. Compare your result to the center of possibilities.

Answer:
  The sorted data appear below.

2	3	3	3	3	3	4	4	4	4
5	5	5	5	5	5	6	6	6	6
6	6	6	7	7	7	7	7	7	7
7	8	8	8	8	8	8	8	9	9
9	9	9	10	10	10	10	10	11	12

Therefrom we get S, whose contents are

$$\begin{array}{ccccc} 2 & 3 & 4 & 5 & 6 & 7 \\ 8 & 9 & 10 & 11 & 12 \end{array}$$

There are 11 values, so

$$\mathcal{C} = \{7\}.$$

The minimum possible value is 2 as each die must have at least one pip on its topmost face. The maximum possible value is 12 as each die can have no more than six pips on its topmost face. The number of pips must be an integer and S contains all integers from 2 to 12, inclusive. We conclude $S = \Omega$ and hence $\mathcal{C} = \mathfrak{C}$.

Example 4:

Determine the center of observations for the two-success data. Compare your result to the center of possibilities.

Answer:

The sorted data appear below.

2	2	2	2	2	2	2	2	2	3
3	3	3	3	3	3	3	3	3	3
3	3	3	3	3	4	4	4	4	4
4	4	4	4	4	4	4	4	4	4
5	5	5	5	5	5	5	5	5	5
5	5	5	5	5	5	5	6	6	6
6	6	6	6	6	6	7	7	7	7
8	8	8	8	9	9	9	9	9	10
10	10	10	10	10	11	11	11	11	12
12	14	14	14	14	14	15	16	17	28

Wherefrom we get S, whose contents is shown below.

$$\begin{array}{cccccccc} 2 & 3 & 4 & 5 & 6 & 7 & 8 & 9 \\ 10 & 11 & 12 & 14 & 15 & 16 & 17 & 28 \end{array}$$

There are 16 distinct values, so

$$\mathcal{C} = \{9, 10\}.$$

5.1. CENTER OF OBSERVATIONS

The population is infinite and has no center.

Median of Observations

SOME PEOPLE DO not like the center of observations because it returns a set. These people prefer to have a single value for the center. The **median of observations** is the median of the set of distinct observations. The median of observations is denoted $\hat{\mathcal{C}}$. If $k = 2m-1$ then
$$\hat{\mathcal{C}} = y_m$$
and if $k = 2m$ then
$$\hat{\mathcal{C}} = \frac{y_m + y_{m+1}}{2}.$$
Thus, we see that it is closely related to the center of observations — which is why we did not make it a separate section.

Example 5:
Determine the median of observations for the two-successes data.

Answer:
 Recall that
$$\mathcal{C} = \{9, 10\}.$$
Therefore,
$$\hat{\mathcal{C}} = \frac{9+10}{2} = \frac{19}{2} = 9.5.$$

Example 6:
Determine the median of observations for the dice roll data.

Answer:

Recall that
$$C = \{7\}.$$
Therefore,
$$\hat{C} = 7.$$

Exercises 5.1

1. One day while walking through a field you notice there seem to be a lot of stones lying on the ground. Curious about what the average size of a stone is in the field you randomly collect 30 stones. You weigh each stone and record its weight in grams. The following table shows the raw data.

110	126	75	110	126	94
90	112	111	139	131	92
102	103	96	75	67	101
104	111	100	65	92	93
106	104	93	115	109	95

   Compare the center of observations to the center of possibilities. Also, determine the median of observations for this data.

2. The scores, in the order graded, on a recent midterm examination in Statistics, a class with 30 students, are given below.

70	80	99	98	85	89
87	79	83	38	69	70
60	69	78	40	75	56
70	51	99	69	95	86
57	53	47	50	55	81

   Compare the center of observations to the center of possibilities. Also, determine the median of observations for this data.

3. Below is a sample of the monthly salary for 30 randomly selected people in mid-management in thousands of dollars.

3.08	3.18	3.13	7.41	3.73	3.05
4.17	3.58	3.36	3.27	3.36	4.74
3.32	3.61	4.26	3.02	3.45	4.06
6.73	4.72	3.13	3.15	3.70	3.59
3.12	3.61	5.03	3.53	3.20	3.32

Compare the center of observations to the center of possibilities. Also, determine the median of observations for this data.

4. A pair of dice, one red and one green, is rolled in the following manner. On the first toss both dice are tossed. Then the red die is tossed while the green die keeps its value to get the second sum. Then the green die is tossed while the red die keeps its value. You continue tossing alternately the red then the green, die and generate the following table.

8	11	8	6	10	9	9
7	2	5	10	11	8	5
3	6	9	7	7	9	10
7	3	3	7	9	8	7
5	4	8	11	11	10	10

Compare the center of observations to the center of possibilities. Also, determine the median of observations for this data.

5. The following random sample of 72 observations comes from a gamma distribution, so $\mu_{min} = 0$ and μ_{max} has no upper limit. The raw data are

2.15	1.92	1.92	2.20	3.67	4.55	3.32	2.71	4.87
2.09	1.13	1.09	2.12	4.89	3.03	3.17	4.74	4.44
4.61	2.78	2.97	2.78	2.88	3.66	2.55	4.40	3.54
2.18	3.43	3.84	4.60	1.76	2.85	2.27	1.38	1.24
4.22	1.67	4.51	3.50	6.69	5.92	6.42	0.95	4.57
3.10	3.26	2.95	2.45	1.95	2.06	3.10	2.03	4.98
0.86	2.61	1.92	2.78	1.01	2.64	3.14	0.86	3.68
2.71	3.06	4.33	4.63	2.47	2.92	1.38	2.81	5.71

Compare the center of observations to the center of possibilities. Also, determine the median of observations for this data.

5.2 Midrange

THE MOST CENTRAL value has more than one interpretation. Last section we discussed the center of observations; that is, the distinct observations. This certainly qualifies for most central considering there are as many distinct values above it as below it. Now, we want to look at another type of most central.

Consider all of the values in the population, from the absolute minimum μ_{min} to the absolute maximum μ_{max}. This forms an interval which contains all possible values. The middle of this interval, denoted $\boldsymbol{\mu_{mid}}$, can be considered most central for the distance from the smallest to the middle is equal to the difference from the middle to the largest. This is the **population midrange**.

$$\mu_{mid} = \frac{\mu_{min} + \mu_{max}}{2}.$$

The problem is that it does not always exist, sometimes because the minimum and/or maximum are not well-defined, sometimes because the maximum and/or minimum is unbounded.

Samples, however, are always finite. Whereupon, x_{min} and x_{max} always exist. The **sample midrange**, denoted $\boldsymbol{x_{mid}}$, is the midpoint between the sample extremes.

$$x_{mid} = \frac{x_{min} + x_{max}}{2}.$$

5.2. MIDRANGE

Example 1:
Determine the midrange for the oyster data and compare it to the midrange of the population.
Answer:
Recall: $x_{\min} = 5.76$ grams and $x_{\max} = 13.18$ grams. Therefore,

$$x_{\text{mid}} = \frac{5.76 + 13.18}{2} = \frac{18.94}{2} = 9.47 \text{ grams}.$$

Neither $\mu_{\min}$ nor $\mu_{\max}$ is well-defined. Therefore, μ_{mid} does not exist.

Example 2:
Determine the midrange for the C++ midterm scores and compare it to the midrange of the population.
Answer:
Recall: $x_{\min} = 46$ and $x_{\max} = 95$. Wherefrom,

$$x_{\text{mid}} = \frac{46 + 95}{2} = \frac{141}{2} = 70.5.$$

It is possible to miss every question, so $\mu_{\min} = 0$. It is also possible to get every question correct, so $\mu_{\max} = 100$. Therefrom,

$$\mu_{\text{mid}} = \frac{0 + 100}{2} = 50.$$

Exercises 5.2

1. One day while walking through a field you notice there seem to be a lot of stones lying on the ground. Curious about what the average

size of a stone is in the field you randomly collect 30 stones. You weigh each stone and record its weight in grams. The following table shows the raw data.

110	126	75	110	126	94
90	112	111	139	131	92
102	103	96	75	67	101
104	111	100	65	92	93
106	104	93	115	109	95

Compare the sample midrange to the population midrange.

2. The scores, in the order graded, on a recent midterm examination in Statistics, a class with 30 students, are given below.

70	80	99	98	85	89
87	79	83	38	69	70
60	69	78	40	75	56
70	51	99	69	95	86
57	53	47	50	55	81

Compare the sample midrange to the population midrange.

3. Below is a sample of the monthly salary for 30 randomly selected people in mid-management in thousands of dollars.

3.08	3.18	3.13	7.41	3.73	3.05
4.17	3.58	3.36	3.27	3.36	4.74
3.32	3.61	4.26	3.02	3.45	4.06
6.73	4.72	3.13	3.15	3.70	3.59
3.12	3.61	5.03	3.53	3.20	3.32

Compare the sample midrange to the population midrange.

4. A pair of dice, one red and one green, is rolled in the following manner. On the first toss both dice are tossed. Then the red die is tossed while the green die keeps its value to get the second sum. Then the green die is tossed while the red die keeps its value. You continue tossing alternately the red then the green, die and generate the following table.

5.3. MEDIAN

8	11	8	6	10	9	9
7	2	5	10	11	8	5
3	6	9	7	7	9	10
7	3	3	7	9	8	7
5	4	8	11	11	10	10

Compare the sample midrange to the population midrange.

5. The following random sample of 72 observations comes from a gamma distribution, so $\mu_{min} = 0$ and μ_{max} has no upper limit. The raw data are

2.15	1.92	1.92	2.20	3.67	4.55	3.32	2.71	4.87
2.09	1.13	1.09	2.12	4.89	3.03	3.17	4.74	4.44
4.61	2.78	2.97	2.78	2.88	3.66	2.55	4.40	3.54
2.18	3.43	3.84	4.60	1.76	2.85	2.27	1.38	1.24
4.22	1.67	4.51	3.50	6.69	5.92	6.42	0.95	4.57
3.10	3.26	2.95	2.45	1.95	2.06	3.10	2.03	4.98
0.86	2.61	1.92	2.78	1.01	2.64	3.14	0.86	3.68
2.71	3.06	4.33	4.63	2.47	2.92	1.38	2.81	5.71

Compare the sample midrange to the population midrange.

5.3 Median

W<small>E DISCUSSED THE</small> median when we talked about box-and-whiskers diagrams. As we have seen, there are several methods used to calculate the median. The **median** is the center of the population in that it corresponds to the 50th percentile rank of the population.

The estimate of effective position for the median using averaging, the nearest, estimation and the median-of-medians methods is

$$\#(M) = \frac{n+1}{2}.$$

If there exists an integer $m = \#(M)$ then these methods all calculate $M = x_{(m)}$, the sorted datum in position m. If, however, this is not an integer, then it lies halfway between integers m and $m+1$. These methods then calculate
$$M = \frac{x_{(m)} + x_{(m+1)}}{2}.$$
When using the even-observation variation of the nearest method, then it will be $x_{(m)}$, when m is even or $x_{(m+1)}$ when m is odd.

The regular proper method calculates $\#(M)$ a little differently. It calculates this as
$$\#(M) = \frac{\#(\mu_{\min}) + \#(\mu_{\max})}{2},$$
which is the same as the methods above when neither $\mu_{\min}$ nor $\mu_{\max}$ is an observation, for then $\#(\mu_{\min}) = 0$ and $\#(\mu_{\max}) = n+1$. This, however, is only the beginning as the regular proper method then sets $m = \lfloor \#(M) \rfloor$ and determines $\#(x_{(m)})$. If $\#(x_{(m)}) = \#(M)$ then $M = x_{(m)}$; otherwise we either have $\#(x_{(m)}) < \#(M)$ or $\#(x_{(m)}) > \#(M)$.

If $\#(x_{(m)}) < \#(M)$ then $m_1 = \#(x_{(m)})$ and $y_1 = x_{(m)}$. We set y_2 to be the next higher distinct observation with $m_2 = \#(y_2)$. If $\#(x_{(m)}) > \#(M)$ then $m_2 = \#(x_{(m)})$ and $y_2 = x_{(m)}$. We set y_1 to be the next lower distinct observation with $m_1 = \#(y_1)$. In either event, we use interpolation between (y_1, m_1) and (y_2, m_2) to determine M as
$$M = \frac{y_1 \cdot [m_2 - \#(M)] + y_2 \cdot [\#(M) - m_1]}{m_2 - m_1}.$$

When we restrict ourselves entirely to the sample, the restricted proper method sets
$$\#(M) = \frac{\#(x_{\min}) + \#(x_{\max})}{2}.$$

5.3. MEDIAN

When both $\mu_{\min}$ and $\mu_{\max}$ are observed values then it produces identical results as the regular proper method. It also produces identical results when $x_{\min}$ and $x_{\max}$ are both unique and either both or neither are population extremes. The details for the interpolation are identical to the regular version.

The traditional sample method uses

$$\#(M) = \frac{n}{2}.$$

If there exists an integer $m = \#(M)$ then

$$M = \frac{x_{(m)} + x_{(m+1)}}{2};$$

otherwise $M = x_{(m+1)}$. In some sense this seems like the opposite of the averaging method — in fact, they are the same and always produce identical results for the median — only the meaning of m is different for odd n. The previous methods count it as $n = 2m - 1$ while this method considers it to be $n = 2m + 1$.

Using grouped data the effective position of the median is

$$\#(M) = \frac{n}{2},$$

just as in the traditional sample method. Now we calculate an offset until

$$\text{offset} = \#(M) - \sum_{j=1}^{i-1} n_j \leq n_i.$$

The value is calculated as

$$M = t_{i-1} + \frac{\text{offset} \cdot (t_i - t_{i-1})}{n_i},$$

where $t_0, t_1, \ldots, t_g$ are the endpoints of the group intervals. Note: The above does not assume that all intervals have the same width. It also assumes that interval i goes from t_{i-1} to t_i and n_i is the number of observations in interval i.

Because the averaging, nearest, estimation, median-of-medians and traditional sample always compute the same value, they can all be considered one method. [The even-observation variation of the nearest method is rarely used in the literature.] The restricted and regular proper methods usually produce slightly different results so we tend to keep them listed as separate methods. Although any time the effective address of the median is the same in the two methods, their medians will also agree and there is no need to do both. Therefore, there are three methods for ungrouped data and one method for grouped data; that is, there are four methods for computing the median.

Example 1:
Determine the median of the two-success data using all four methods. For the grouped methods, use five and seven intervals — don't worry about the empty interval in this as it will not affect the median calculation.

Answer:
The frequency table appears below.

2: 9	4: 15	6: 9	8: 4	10: 6	12: 2	15: 1	17: 1
3: 16	5: 17	7: 4	9: 5	11: 4	14: 5	16: 1	28: 1

The averaging, nearest, estimation and median-of-medians method calculate $\#(M) = 101/2 = 50.5$. The traditional sample method calculates it as $\#(M) = 100/2 = 50$, but in all five cases

$$M = \frac{x_{(50)} + x_{(51)}}{2} = \frac{5+5}{2} = 5.$$

5.3. MEDIAN

The regular proper method has $\#(\mu_{\min}) = \#(x_{\min}) = (1+9)/2 = 5$ and $\#(\mu_{\max}) = 101$; thus,

$$\#(M) = \frac{5+101}{2} = 53.$$

Now, $x_{(53)} = 5$, but $\#(5) = (41+57)/2 = 49 < 53$; wherefrom, $y_1 = 5$, $m_1 = 49$, $y_2 = 6$ and $m_2 = (58+66)/2 = 62$. Whence,

$$M = \frac{5 \cdot (62-53) + 6 \cdot (53-49)}{62-49} = \frac{45+24}{13} = \frac{69}{13} \approx 5.31.$$

The restricted proper method changes $\mu_{\max}$ to $x_{\max}$, the lower bound being the same as $\mu_{\min} = x_{\min}$. Now, $\#(x_{\max}) = 100$ as 28 is unique. Therefrom,

$$\#(M) = \frac{5+100}{2} = \frac{105}{2} = 52.5.$$

The same (y_1, m_1) and (y_2, m_2) are used in the interpolation — only $\#(M)$ is different. Thereby,

$$M = \frac{5 \cdot (62-52.5) + 6 \cdot (52.5-49)}{62-49} = \frac{47.5+21}{13} = \frac{137}{26} \approx 5.27.$$

With five intervals, the common width is

$$w = \frac{28-2}{5} = 5.2.$$

This gives the following frequency table.

I_i	m_i	f_i
$[2, 7.2)$	4.6	70
$[7.2, 12.4)$	9.8	21
$[12.4, 17.6)$	15.0	8
$[17.6, 22.8)$	20.2	0
$[22.8, 28]$	25.4	1

The group method calculates $\#(M) = 100/2 = 50$, so the median lies in the first interval. We conclude

$$M = 2 + \frac{50 \cdot (7.2 - 2)}{70} = \frac{40}{7} \approx 5.71.$$

With seven intervals, the common width is

$$w = \frac{28 - 2}{7} = \frac{26}{7} \approx 3.714.$$

We round this up to the nearest 0.01, so $w = 3.72$. This means, for the purpose of generating t_1 through t_7,

$$t_0 = \frac{2 + 28 - 7 \cdot 3.72}{2} = 1.98.$$

In reality, as $\mu_{\min} = 2$, $t_0 = 2$ and the first interval is slightly smaller than the others. The frequency table for this is

I_i	m_i	f_i
$[2.00, 5.70)$	3.85	57
$[5.70, 9.42)$	7.56	22
$[9.42, 13.14)$	11.28	12
$[13.14, 16.86)$	15.00	7
$[16.86, 20.58)$	18.72	1
$[20.58, 24.30)$	22.44	0
$[24.30, 28.02]$	26.16	1

The effective position is the same as we are still using the group method, so the median is still within the first interval. Whereupon we determine

$$M = 2 + \frac{50 \cdot (5.7 - 2)}{57} = \frac{299}{57} \approx 5.25.$$

Example 2:
Determine the median for the random gamma data introduced in

5.3. MEDIAN

Example 1, §3.3, which started on page 63 using all four methods. For the grouped method use four and five intervals.

Answer:
The frequency table is repeated below.

4: 1	9: 2	13: 3	17: 2	24: 2
5: 4	10: 1	14: 2	18: 1	27: 1
8: 1	12: 1	15: 1	22: 2	31: 1

The averaging, nearest, estimation and median-of-medians method calculate $\#(M) = 26/2 = 13$. The traditional sample method calculates it as $\#(M) = 25/2 = 12.5$, but in all five cases we conclude

$$M = x_{(13)} = 13.$$

Neither $\mu_{\min} = 0$ nor $\mu_{\max}$ (who has no finite upper limit) is present; thus, $\#(\mu_{\min}) = 0$ and $\#(\mu_{\max}) = 26$. We conclude by the regular proper method that $\#(M) = (0 + 26)/2 = 13$. Now, $x_{\min} = 4$ and $x_{\max} = 31$ are unique, so the restricted proper method also gets $\#(M) = 13$ because $1 + 25 = 0 + 26$. Therefore, the regular and restricted proper methods will agree.

We note that $\#(x_{(13)}) = (11 + 13)/2 = 12 < 13$, so $y_1 = 13$ and $m_1 = 12$. The next higher observation is $y_2 = x_{(14)} = 14$ with $m_2 = (14 + 15)/2 = 14.5$. By interpolation we obtain

$$M = \frac{13 \cdot (14.5 - 13) + 14 \cdot (13 - 12)}{14.5 - 12} = \frac{19.5 + 14}{2.5} = \frac{67}{5} = 13.4.$$

With four intervals, the common width is

$$w = \frac{31 - 4}{4} = 6.75.$$

This is not bad, but we will round this up to the nearest 0.1 to get $w = 6.8$. From this,

$$t_0 = \frac{4 + 31 - 4(6.8)}{2} = 3.9.$$

This gives the following frequency table

I_i	$[3.9, 10.7)$	$[10.7, 17.5)$	$[17.5, 24.3)$	$[24.3, 31.1]$
m_i	7.3	14.1	20.9	27.7
f_i	9	9	5	2

The group method calculates $\#(M) = 25/2 = 12.5$, which makes the median lie in the second interval. The offset into this interval is

$$\text{offset} = 12.5 - 9 = 3.5 < 9.$$

From which we get

$$M = 10.7 + \frac{3.5 \cdot (17.5 - 10.7)}{9} = \frac{1201}{90} \approx 13.344.$$

With five intervals, the common width is

$$w = \frac{31 - 4}{5} = 5.4.$$

We'll keep this, so $t_0 = 4$. We get the following frequency table.

I_i	$[4.0, 9.4)$	$[9.4, 14.8)$	$[14.8, 20.2)$	$[20.2, 25.6)$	$[25.6, 31.0]$
m_i	6.7	12.1	17.5	22.9	28.3
f_i	8	7	4	4	2

The group method is still being used so $\#(M) = 12.5$ and the median is still in the second interval. The offset is now

$$\text{offset} = 12.5 - 8 = 4.5 < 7.$$

From this we obtain

$$M = 9.4 + \frac{4.5 \cdot (14.8 - 9.4)}{7} = \frac{901}{70} \approx 12.871.$$

5.3. MEDIAN

Exercises 5.3

1. One day while walking through a field you notice there seem to be a lot of stones lying on the ground. Curious about what the average size of a stone is in the field you randomly collect 30 stones. You weigh each stone and record its weight in grams. The following table shows the raw data.

110	126	75	110	126	94
90	112	111	139	131	92
102	103	96	75	67	101
104	111	100	65	92	93
106	104	93	115	109	95

   Compute the median using all four methods. For the grouped method use five intervals.

2. The scores, in the order graded, on a recent midterm examination in Statistics, a class with 30 students, are given below.

70	80	99	98	85	89
87	79	83	38	69	70
60	69	78	40	75	56
70	51	99	69	95	86
57	53	47	50	55	81

   Compute the median using all four methods. For the grouped method use five intervals.

3. Below is a sample of the monthly salary for 30 randomly selected people in mid-management in thousands of dollars.

3.08	3.18	3.13	7.41	3.73	3.05
4.17	3.58	3.36	3.27	3.36	4.74
3.32	3.61	4.26	3.02	3.45	4.06
6.73	4.72	3.13	3.15	3.70	3.59
3.12	3.61	5.03	3.53	3.20	3.32

Compute the median using all four methods. For the grouped method use five intervals.

4. A pair of dice, one red and one green, is rolled in the following manner. On the first toss both dice are tossed. Then the red die is tossed while the green die keeps its value to get the second sum. Then the green die is tossed while the red die keeps its value. You continue tossing alternately the red then the green, die and generate the following table.

8	11	8	6	10	9	9
7	2	5	10	11	8	5
3	6	9	7	7	9	10
7	3	3	7	9	8	7
5	4	8	11	11	10	10

Compute the median using all four methods. For the grouped method use five intervals.

5. The following random sample of 72 observations comes from a gamma distribution, so $\mu_{min} = 0$ and μ_{max} has no upper limit. The raw data are

2.15	1.92	1.92	2.20	3.67	4.55	3.32	2.71	4.87
2.09	1.13	1.09	2.12	4.89	3.03	3.17	4.74	4.44
4.61	2.78	2.97	2.78	2.88	3.66	2.55	4.40	3.54
2.18	3.43	3.84	4.60	1.76	2.85	2.27	1.38	1.24
4.22	1.67	4.51	3.50	6.69	5.92	6.42	0.95	4.57
3.10	3.26	2.95	2.45	1.95	2.06	3.10	2.03	4.98
0.86	2.61	1.92	2.78	1.01	2.64	3.14	0.86	3.68
2.71	3.06	4.33	4.63	2.47	2.92	1.38	2.81	5.71

Compute the median using all four methods. For the grouped method use five and six intervals.

5.4 Walsh Sum Median

W<small>HEN RESTRICTING</small> oneself to the data, especially with data which are summable, the median does not take the interaction effects into account. There is a modification of the median which does try to take some of these interactions into account. That will be the subject of this section.

Suppose $x_{(i)}$ and $x_{(j)}$ are two observations, not necessarily distinct. The **Walsh sum of $x_{(i)}$ and $x_{(j)}$**, denoted $\boldsymbol{W(i,j)}$, is defined as

$$W(i,j) = \frac{x_{(i)} + x_{(j)}}{2}.$$

It is assumed that $1 \leq i \leq j \leq n$, so there are

$$\frac{n \cdot (n+1)}{2}$$

Walsh sums.

In the two-success data there are 100 observations. This means there are

$$\frac{100 \cdot 101}{2} = 5050$$

Walsh sums. However, there are only 16 distinct observed values. This means the number of *distinct* Walsh sums cannot be greater than

$$\frac{16 \cdot 17}{2} = 136.$$

It is quite possible for $x_{(i)} + x_{(j)} = x_{(k)} + x_{(\ell)}$ even though none of the values equal each other. Suppose $x_{(1)} = 1$, $x_{(2)} = 3$, $x_{(3)} = 4$ and $x_{(4)} = 6$ in a tiny sample of size four. We can see that

$$1 + 6 = 3 + 4,$$

yet none of the addends are equal to each other. Something similar could certainly occur with the distinct values of a sample such as the two-success data. This is why we mentioned the number of distinct Walsh sums cannot be greater than 136 in the two-success data.

It is actually not difficult to determine the frequency of $W(i,j)$, where we now use

$$S = \{y_1, \ldots, y_k\},$$

the set of distinct observations. We assume $\text{frq}(y_i) = f_i$ for $i = 1, \ldots, k$. The formula is $(i \leq j)$

$$\text{frq}(W(i,j)) = \begin{cases} \dfrac{f_i \cdot (f_i + 1)}{2} & i = j \\ f_i \cdot f_j & i < j \end{cases}.$$

We consider the collection of all Walsh sums, with their frequencies, as being a sample of size $n(n+1)/2$ and determine the median of this sample to get the **Walsh sum median**.

Example 1:
Determine the Walsh sum median for the two-success data.

Answer:
The sorted data appear below.

2	2	2	2	2	2	2	2	2	3	3	3	3	3	3
3	3	3	3	3	3	3	3	3	3	4	4	4	4	4
4	4	4	4	4	4	4	4	4	4	5	5	5	5	5
5	5	5	5	5	5	5	5	5	5	5	5	6	6	6
6	6	6	6	6	6	7	7	7	7	8	8	8	8	9
9	9	9	9	10	10	10	10	10	10	11	11	11	11	12
12	14	14	14	14	14	15	16	17	28					

Which gives the following tally table.

2: 9	4: 15	6: 9	8: 4	10: 6	12: 2	15: 1	17: 1
3: 16	5: 17	7: 4	9: 5	11: 4	14: 5	16: 1	28: 1

5.4. WALSH SUM MEDIAN

The Walsh sums based upon this are:

W	2	3	4	5	6	7	8	9	10	11	12	14	15	16	17	28
2	2	2.5	3.0	3.5	4.0	4.5	5.0	5.5	6.0	6.5	7.0	8.0	8.5	9.0	9.5	15.0
3		3.0	3.5	4.0	4.5	5.0	5.5	6.0	6.5	7.0	7.5	8.5	9.0	9.5	10.0	15.5
4			4.0	4.5	5.0	5.5	6.0	6.5	7.0	7.5	8.0	9.0	9.5	10.0	10.5	16.0
5				5.0	5.5	6.0	6.5	7.0	7.5	8.0	8.5	9.5	10.0	10.5	11.0	16.5
6					6.0	6.5	7.0	7.5	8.0	8.5	9.0	10.0	10.5	11.0	11.5	17.0
7						7.0	7.5	8.0	8.5	9.0	9.5	10.5	11.0	11.5	12.0	17.5
8							8.0	8.5	9.0	9.5	10.0	11.0	11.5	12.0	12.5	18.0
9								9.0	9.5	10.0	10.5	11.5	12.0	12.5	13.0	18.5
10									10.0	10.5	11.0	12.0	12.5	13.0	13.5	19.0
11										11.0	11.5	12.5	13.0	13.5	14.0	19.5
12											12.0	13.0	13.5	14.0	14.5	20.0
14												14.0	14.5	15.0	15.5	21.0
15													15.0	15.5	16.0	21.5
16														16.0	16.5	22.0
17															17.0	22.5
28																28.0

The frequency for each of these cases is

f	2	3	4	5	6	7	8	9	10	11	12	14	15	16	17	28
2	45	144	135	153	81	36	36	45	54	36	18	45	9	9	9	9
3		136	240	272	144	64	64	80	96	64	32	80	16	16	16	16
4			120	255	135	60	60	75	90	60	30	75	15	15	15	15
5				153	153	68	68	85	102	68	34	85	17	17	17	17
6					45	36	36	45	54	36	18	45	9	9	9	9
7						10	16	20	24	16	8	20	4	4	4	4
8							10	20	24	16	8	20	4	4	4	4
9								15	30	20	10	25	5	5	5	5
10									21	24	12	30	6	6	6	6
11										10	8	20	4	4	4	4
12											3	10	2	2	2	2
14												15	5	5	5	5
15													1	1	1	1
16														1	1	1
17															1	1
28																1

Combining Walsh sums which are equal we get the following frequency table — there are only 42 distinct Walsh sums.

2.0: 45	5.0: 388	8.0: 227	11.0: 72	14.0: 21	17.0: 10	20.0: 2
2.5: 144	5.5: 322	8.5: 203	11.5: 50	14.5: 7	17.5: 4	21.0: 5
3.0: 271	6.0: 307	9.0: 173	12.0: 46	15.0: 15	18.0: 4	21.5: 1
3.5: 393	6.5: 311	9.5: 179	12.5: 35	15.5: 22	18.5: 5	22.0: 1
4.0: 473	7.0: 303	10.0: 142	13.0: 25	16.0: 17	19.0: 6	22.5: 1
4.5: 435	7.5: 255	10.5: 95	13.5: 12	16.5: 18	19.5: 4	28.0: 1

We leave it to you to check that these add up to $100 \cdot 101/2 = 5050$, the total number of Walsh sums for the data.

The traditional sample method calculates

$$\#(M) = \frac{5050}{2} = 2525$$

Both $w_{(2525)}$ and $w_{(2526)} = 6.0$, so $M = 6.0$.

The restricted proper method has

$$\#(2.0) = \frac{1 + 45}{2} = 23 \quad \text{and} \quad \#(28.0) = 5050.$$

From which

$$\#(M) = \frac{23 + 5050}{2} = 2536.5.$$

Now, $w_{(2536)} = 6.0$ and $\#(6.0) = (2472 + 2778)/2 = 2625 > 2536.5$; whence, $y_2 = 6.0$, $m_2 = 2625$, $y_1 = 5.5$ and $m_1 = (2150 + 2471)/2 = 2310.5$. The median is

$$M = \frac{5.5 \cdot (2625 - 2536.5) + 6.0 \cdot (2536.5 - 2310.5)}{2625 - 2310.5}$$
$$= \frac{486.75 + 1356}{314.5} = \frac{7371}{1258} \approx 5.8593.$$

Many statistics software packages have methods for calculating the Walsh sum median

5.4. WALSH SUM MEDIAN

Example 2:
Determine the Walsh sum median for the dice roll data.

Answer:
The frequency table for the dice roll data is

2: 1	5: 6	8: 7	11: 1
3: 5	6: 7	9: 5	12: 1
4: 4	7: 8	10: 5	

Below we show all Walsh sums and their frequencies

Wf	2	3	4	5	6	7	8	9	10	11	12
2	2 1	2.5 5	3.0 4	3.5 6	4.0 7	4.5 8	5.0 7	5.5 5	6.0 5	6.5 1	7.0 1
3		3.0 15	3.5 20	4.0 30	4.5 35	5.0 40	5.5 35	6.0 25	6.5 25	7.0 5	7.5 5
4			4.0 10	4.5 24	5.0 28	5.5 32	6.0 28	6.5 20	7.0 20	7.5 4	8.0 4
5				5.0 21	5.5 42	6.0 48	6.5 42	7.0 30	7.5 30	8.0 6	8.5 6
6					6.0 28	6.5 56	7.0 49	7.5 35	8.0 35	8.5 7	9.0 7
7						7.0 36	7.5 56	8.0 40	8.5 40	9.0 8	9.5 8
8							8.0 28	8.5 35	9.0 35	9.5 7	10.0 7
9								9.0 15	9.5 25	10.0 5	10.5 5
10									10.0 15	10.5 5	11.0 5
11										11.0 1	11.5 1
12											12.0 1

Combining the Walsh sums which are equal, we get the following frequency table

2.0: 1	3.5: 26	5.0: 96	6.5: 144	8.0: 113	9.5: 40	11.0: 6
2.5: 5	4.0: 47	5.5: 114	7.0: 141	8.5: 88	10.0: 27	11.5: 1
3.0: 19	4.5: 67	6.0: 134	7.5: 130	9.0: 65	10.5: 10	12.0: 1

We leave it to you to verify they add up to $50 \cdot 51/2 = 1275$.
The traditional sample method calculates

$$\#(M) = \frac{1275}{2} = 637.5,$$

which makes $M = w_{(638)} = 6.5$.

The restricted proper method notes that $w_{(1)} = 2$ and $w_{(1275)} = 12$ are unique; therefore,

$$\#(M) = \frac{1 + 1275}{2} = 638.$$

Now, $w_{(638)} - 6.5$, as noted before, but $\#(6.5) = (510 + 653)/2 = 581.5 < 638$; thus, $y_1 = 6.5$, $m_1 = 581.5$, $y_2 = 7.0$ and $m_2 = (654 + 794)/2 = 724$. By interpolation,

$$M = \frac{6.5 \cdot (724 - 638) + 7.0 \cdot (638 - 581.5)}{724 - 581.5}$$
$$= \frac{559 + 395.5}{142.5} \frac{1909}{285} \approx 6.6982.$$

Walsh Sum Central Median

SOME PEOPLE disregard the frequency of the distinct Walsh sums and take the median of the generated values, treating each as unique. This is a straightforward modification of the median of observations. There is no standard notation for this, but we will use $\hat{W}$ to denote the **Walsh sum central median**.

Example 3:
Determine the Walsh sum central median for the two-success data.

Answer:
The set of distinct Walsh sums contains

5.4. WALSH SUM MEDIAN

2.0	2.5	3.0	3.5	4.0	4.5	5.0
5.5	6.0	6.5	7.0	7.5	8.0	8.5
9.0	9.5	10.0	10.5	11.0	11.5	12.0
12.5	13.0	13.5	14.0	14.5	15.0	15.5
16.0	16.5	17.0	17.5	18.0	18.5	19.0
19.5	20.0	21.0	21.5	22.0	22.5	28.0

There are 42 values; hence

$$\hat{W} = \frac{12 + 12.5}{2} = 12.25.$$

Example 4:
Determine the Walsh sum central median for the dice roll data.

Answer:
The set of distinct Walsh sums contains

2.0	2.5	3.0	3.5	4.0	4.5	5.0
5.5	6.0	6.5	7.0	7.5	8.0	8.5
9.0	9.5	10.0	10.5	11.0	11.5	12.0

There are 21 values; hence

$$\hat{W} = 7.0.$$

Grouped Data

THERE IS A modification of this method which can be done with grouped data. The midpoint of each interval is used in place of the observation and the frequency of that midpoint is the number of observations in that interval. If any interval is empty, then that interval is not used at all in calculating Walsh sums.

Example 5:
Group the two-success data into seven intervals and determine the Walsh sum median by the traditional sample and the restricted proper methods. In addition, determine the Walsh sum central median.

Answer:
The common width of each interval should be

$$w = \frac{28 - 2}{7} = \frac{26}{7} \approx 3.7143.$$

We will round this up to the nearest 0.1 and make $w = 3.8$. For determining the rest of the endpoints we use

$$t_0 = \frac{2 + 28 - 7 \cdot 3.8}{2} = 1.7,$$

but realize that $\mu_{\min} = 2$, so the first interval will be shorter. The frequency table for the intervals is

I_i	m_i	f_i
$[2, 5.5)$	3.75	57
$[5.5, 9.3)$	7.40	22
$[9.3, 13.1)$	11.20	12
$[13.1, 16.9)$	15.00	7
$[16.9, 20.7)$	18.80	1
$[20.7, 24.5)$	22.60	0
$[24.5, 28.3]$	26.40	1

This give the following frequencies and values for use in the Walsh sums

3.75: 57   11.2: 12   18.8: 1
7.40: 22   15.0: 7   26.4: 1

From the above we get the following Walsh sums and their frequencies

5.4. WALSH SUM MEDIAN

Wf	3.75	7.4	11.2	15	18.8	26.4
3.75	3.75	5.575	7.475	9.375	11.275	15.075
	1653	1254	684	399	57	57
7.4		7.400	9.300	11.200	13.100	16.900
		253	264	154	22	22
11.2			11.200	13.100	15.000	18.800
			78	84	12	12
15				15.000	16.900	20.700
				28	7	7
18.8					18.800	22.600
					1	1
26.4						26.400
						1

The distinct Walsh sums and their frequencies are

    3.750: 1653   9.300: 264   13.100: 106   18.800: 13
    5.575: 1254   9.375: 399   15.000: 40   20.700: 7
    7.400: 253   11.200: 232   15.075: 57   22.600: 1
    7.475: 684   11.275: 57   16.900: 29   26.400: 1

There a 16 distinct Walsh sums, so the Walsh sum central median is $\hat{W} = (11.275 + 13.1)/2 = 12.1875$.

There are 5050 Walsh sums, with repetition, so the traditional sample method has
$$\#(M) = \frac{5050}{2} = 2525$$
and
$$M = \frac{w_{(2525)} + w_{(2526)}}{2} = \frac{5.575 + 5.575}{2} = 5.575.$$

The restricted proper method notes that
$$\#(w_{\min}) = \frac{1 + 1653}{2} = 827 \quad \text{and} \quad \#(w_{\max}) = 5050.$$

Therefrom,
$$\#(M) = \frac{827 + 5050}{2} = 2938.5.$$

From the above tally table, $w_{(2938)} = 7.4$ and $\#(7.4) = (2908 + 3160)/2 = 3034 > 2938.5$. We conclude $y_2 = 7.4$, $m_2 = 3034$, $y_1 = 5.575$ and $m_1 =$

$(1654 + 2907)/2 = 2280.5$. By interpolation,

$$M = \frac{5.575 \cdot (3034 - 2938.5) + 7.4 \cdot (2938.5 - 2280.5)}{3034 - 2280.5}$$
$$= \frac{532.4125 + 4869.2}{753.5} = \frac{432129}{60280} \approx 7.1687.$$

Exercises 5.4

1. One day while walking through a field you notice there seem to be a lot of stones lying on the ground. Curious about what the average size of a stone is in the field you randomly collect 30 stones. You weigh each stone and record its weight in grams. The following table shows the raw data.

110	126	75	110	126	94
90	112	111	139	131	92
102	103	96	75	67	101
104	111	100	65	92	93
106	104	93	115	109	95

    Determine the Walsh sum median and Walsh sum central median for this data. If there are more than 16 distinct data values, then use the grouped method to determine the two Walsh sum medians.

2. The scores, in the order graded, on a recent midterm examination in Statistics, a class with 30 students, are given below.

70	80	99	98	85	89
87	79	83	38	69	70
60	69	78	40	75	56
70	51	99	69	95	86
57	53	47	50	55	81

5.4. WALSH SUM MEDIAN

Determine the Walsh sum median and Walsh sum central median for this data. If there are more than 16 distinct data values, then use the grouped method to determine the two Walsh sum medians.

3. Below is a sample of the monthly salary for 30 randomly selected people in mid-management in thousands of dollars.

3.08	3.18	3.13	7.41	3.73	3.05
4.17	3.58	3.36	3.27	3.36	4.74
3.32	3.61	4.26	3.02	3.45	4.06
6.73	4.72	3.13	3.15	3.70	3.59
3.12	3.61	5.03	3.53	3.20	3.32

Determine the Walsh sum median and Walsh sum central median for this data. If there are more than 16 distinct data values, then use the grouped method to determine the two Walsh sum medians.

4. A pair of dice, one red and one green, is rolled in the following manner. On the first toss both dice are tossed. Then the red die is tossed while the green die keeps its value to get the second sum. Then the green die is tossed while the red die keeps its value. You continue tossing alternately the red then the green, die and generate the following table.

8	11	8	6	10	9	9
7	2	5	10	11	8	5
3	6	9	7	7	9	10
7	3	3	7	9	8	7
5	4	8	11	11	10	10

Determine the Walsh sum median and Walsh sum central median for this data. If there are more than 16 distinct data values, then use the grouped method to determine the two Walsh sum medians.

5. The following random sample of 72 observations comes from a gamma distribution, so $\mu_{min} = 0$ and μ_{max} has no upper limit. The raw data are

2.15	1.92	1.92	2.20	3.67	4.55	3.32	2.71	4.87
2.09	1.13	1.09	2.12	4.89	3.03	3.17	4.74	4.44
4.61	2.78	2.97	2.78	2.88	3.66	2.55	4.40	3.54
2.18	3.43	3.84	4.60	1.76	2.85	2.27	1.38	1.24
4.22	1.67	4.51	3.50	6.69	5.92	6.42	0.95	4.57
3.10	3.26	2.95	2.45	1.95	2.06	3.10	2.03	4.98
0.86	2.61	1.92	2.78	1.01	2.64	3.14	0.86	3.68
2.71	3.06	4.33	4.63	2.47	2.92	1.38	2.81	5.71

Determine the Walsh sum median and Walsh sum central median for this data. If there are more than 16 distinct data values, then use the grouped method to determine the two Walsh sum medians.

5.5 Mode

HAVING LOOKED at the most central value, we now turn our attention to the most common value. Why this is considered a problem for centrality is the interpretation of the word central as meaning important. If you are undertaking a decision you need to know what is the most likely outcome to be able to assess whether you want to try it. This is *especially* true if there is any sort of risk involved. There is one statistic for this; which we cover now.

The **mode**, denoted M^o, is the set of most common values. The reason we consider it a set is that often more than one value is tied with the highest frequency. It makes no sense to combine them to create a "mode," for you might end up with a most-common-value that cannot possibly occur — hardly what you could call most common. Thus, the mode is the *set* of most common values.

If the mode is unique then the data are **unimodal**; otherwise, they are **polymodal** (or **multimodal**). Actually, the first few numbers of modes are so common that we have names for them. These

5.5. MODE

names are given in Table 5.1. Although M^o is a set, it is common to omit the braces when the data are unimodal.

Table 5.1: Common modality names.

number of modes	name applied
1	unimodal
> 1	polymodal (multimodal)
2	bimodal
3	trimodal
4	quadrimodal
5	pentamodal

Example 1:
Determine the mode for the two-success data and classify the modality using the names in Table 5.1.

Answer:
The frequency table for this data is repeated below.

2: 9	6: 9	10: 6	15: 1
3: 16	7: 4	11: 4	16: 1
4: 15	8: 4	12: 2	17: 1
5: 17	9: 5	14: 5	28: 1

Wherefrom we see that the highest frequency is frq(5) = 17, so $M^o = \{5\} = 5$ and the data are unimodal. You may notice that frq(3) = 16 and frq(4) = 15. This indicates that were the experiment repeated the data might not be unimodal next time and could potentially even be trimodal.

Example 2:
Determine the mode of the dice-roll data.

Answer:
The frequency table for this data is repeated below.

$$\begin{array}{lll} 2\colon 1 & 5\colon 6 & 8\colon 7 \quad 11\colon 1 \\ 3\colon 5 & 6\colon 7 & 9\colon 5 \quad 12\colon 1 \\ 4\colon 4 & 7\colon 8 & 10\colon 5 \end{array}$$

Therefrom we see that the highest frequency is frq(7) = 8, so $M^o = \{7\} = 7$ and the data are unimodal. Both frq(6) = 7 and frq(8) = 7; also, frq(5) = 6. This indicates that were the experiment repeated the data might not be unimodal next time and could potentially even be quadrimodal.

Of course there is nothing which says the data must be numerical. Categorical data is fine. In some sense it is even better because you are less likely to even *want* to combine the modes.

Example 3:
Determine the mode for the fuþorc data and classify the modality using the names in Table 5.1.

Answer:
The frequency table is listed below.

$$\begin{array}{lllllll}
ᚠ\colon 2 & ᚴ\colon 2 & ᛁ\colon 6 & ᚼ\colon 4 & ᚱ\colon 2 & ᚡ\colon 1 & ᛅ\colon 2 \\
ᚾ\colon 3 & ᚷ\colon 2 & ᚦ\colon 5 & ᛏ\colon 4 & ᛪ\colon 4 & ᛑ\colon 4 & ᛘ\colon 4 \\
ᚣ\colon 3 & ᛈ\colon 2 & ᛋ\colon 1 & ᛒ\colon 4 & ᛖ\colon 2 & ⁎\colon 2 & ᛉ\colon 3 \\
ᚿ\colon 3 & ᚺ\colon 4 & ᛕ\colon 2 & ᛗ\colon 1 & ᚨ\colon 6 & ᛐ\colon 2 & \\
ᚱ\colon 6 & ᛏ\colon 6 & ᚤ\colon 3 & ᛙ\colon 4 & ᚡ\colon 3 & ᛌ\colon 3 &
\end{array}$$

The highest frequency is 6. Specifically, frq(ᚱ) = frq(ᛏ) = frq(ᛁ) = frq(ᚨ) = 6. This makes
$$M^o = \{ᚱ, ᛏ, ᛁ, ᚨ\}$$

5.5. MODE

and the data are quadrimodal. You might notice that frq($\blacklozenge$) = 5 and many had frequency 4. Were we to repeat the experiment then the number of modes might be different — perhaps more, perhaps less.

Exercises 5.5

1. One day while walking through a field you notice there seem to be a lot of stones lying on the ground. Curious about what the average size of a stone is in the field you randomly collect 30 stones. You weigh each stone and record its weight in grams. The following table shows the raw data.

110	126	75	110	126	94
90	112	111	139	131	92
102	103	96	75	67	101
104	111	100	65	92	93
106	104	93	115	109	95

   Determine the mode of the data and classify the modality using the names in Table 5.1.

2. The scores, in the order graded, on a recent midterm examination in Statistics, a class with 30 students, are given below.

70	80	99	98	85	89
87	79	83	38	69	70
60	69	78	40	75	56
70	51	99	69	95	86
57	53	47	50	55	81

   Determine the mode of the data and classify the modality using the names in Table 5.1.

3. Below is a sample of the monthly salary for 30 randomly selected people in mid-management in thousands of dollars.

3.08	3.18	3.13	7.41	3.73	3.05
4.17	3.58	3.36	3.27	3.36	4.74
3.32	3.61	4.26	3.02	3.45	4.06
6.73	4.72	3.13	3.15	3.70	3.59
3.12	3.61	5.03	3.53	3.20	3.32

Determine the mode of the data and classify the modality using the names in Table 5.1.

4. A pair of dice, one red and one green, is rolled in the following manner. On the first toss both dice are tossed. Then the red die is tossed while the green die keeps its value to get the second sum. Then the green die is tossed while the red die keeps its value. You continue tossing alternately the red then the green, die and generate the following table.

8	11	8	6	10	9	9
7	2	5	10	11	8	5
3	6	9	7	7	9	10
7	3	3	7	9	8	7
5	4	8	11	11	10	10

Determine the mode of the data and classify the modality using the names in Table 5.1.

5. The following random sample of 72 observations comes from a gamma distribution, so $\mu_{min} = 0$ and μ_{max} has no upper limit. The raw data are

2.15	1.92	1.92	2.20	3.67	4.55	3.32	2.71	4.87
2.09	1.13	1.09	2.12	4.89	3.03	3.17	4.74	4.44
4.61	2.78	2.97	2.78	2.88	3.66	2.55	4.40	3.54
2.18	3.43	3.84	4.60	1.76	2.85	2.27	1.38	1.24
4.22	1.67	4.51	3.50	6.69	5.92	6.42	0.95	4.57
3.10	3.26	2.95	2.45	1.95	2.06	3.10	2.03	4.98
0.86	2.61	1.92	2.78	1.01	2.64	3.14	0.86	3.68
2.71	3.06	4.33	4.63	2.47	2.92	1.38	2.81	5.71

5.5. MODE

Determine the mode of the data and classify the modality using the names in Table 5.1.

6. A bag of candy-coated chocolate drops has 42 pieces of candy in it. You recorded the colors of each piece as it was removed from the bag and the result is given in the follow table.

red	brown	yellow	brown	green	yellow	yellow
brown	brown	brown	brown	red	brown	red
brown	brown	blue	brown	yellow	brown	blue
green	brown	yellow	yellow	brown	brown	brown
yellow	orange	orange	yellow	yellow	red	brown
brown	blue	red	blue	yellow	yellow	red

Determine the mode of the data and classify the modality using the names in Table 5.1.

7. You toss a coin 20 times and record whether the obverse side (heads) or reverse side (tails) fell up. The result of these twenty tosses is shown below. In the table, H means heads and T means tails.

T	H	T	T	H
H	T	T	H	T
H	H	T	T	H
T	H	H	T	H

Determine the mode of the data and classify the modality using the names in Table 5.1.

8. The elder fuþark consisted of 24 runes, ordered as follows:

ᚠᚢᚦᚨᚱᚲᚷᚹᚺᚾᛁᛃᛇᛈᛉᛊᛏᛒᛖᛗᛚᛜᛟᛞ

You can buy sets of these made of some kind of hard plastic with a bag as they are often used in fortunetelling. They usually come with a blank tile — there is no historical evidence for this blank "rune." Suppose you bought a set. You shake the bag to mix them up, blindly reach in to get one, record the letter and return it to the

bag, mixing it up again. You repeat this process 60 times and obtain
the following data

ᛄ	ᛒ	ᛯ	ᛋ	ᛋ	ᛦ	ᛗ	ᚱ	ᛏ	ᚹ
ᛜ	ᛗ	ᛏ	ᛈ	ᛈ	ᛗ	ᚺ	ᛏ	ᛗ	ᛋ
ᛏ	ᛒ	ᚺ	ᚠ	ᛗ	ᛋ	ᛯ	ᛗ	ᛜ	ᛒ
ᚠ	ᛕ	ᛉ	ᛜ	ᚲ	ᛕ	ᛕ	ᚲ	ᚾ	ᚦ
ᚹ	᛭	ᛋ	ᛜ	ᛁ	ᛏ	ᚾ	ᚠ	ᚱ	ᚱ
ᛄ	ᚹ	ᚠ	ᚲ	ᚹ	ᛉ	ᛄ	ᛉ	ᚺ	ᚠ

The sorted order is

ᚹ	ᚹ	ᚹ	ᚹ	ᚾ	ᚾ	ᚦ	ᚠ	ᚠ	ᚠ
ᚠ	ᚠ	ᚱ	ᚲ	ᚲ	ᚲ	ᛉ	ᛉ	ᛉ	ᛈ
ᛈ	ᚺ	ᚺ	ᚺ	᛭	ᛁ	ᛋ	ᛄ	ᛄ	ᛄ
ᛕ	ᛕ	ᛕ	ᛦ	ᛋ	ᛋ	ᛋ	ᛋ	ᛏ	ᛏ
ᛏ	ᛏ	ᛏ	ᛒ	ᛒ	ᛒ	ᛗ	ᛗ	ᛗ	ᚱ
ᚱ	ᛜ	ᛜ	ᛜ	ᛜ	ᛗ	ᛗ	ᛗ	ᛯ	ᛯ

Determine the mode of the data and classify the modality using the
names in Table 5.1.

5.6 Types of Averages

GIVEN NUMERIC DATA, the way one should take an average depends
upon the nature of the data. In this section we will discuss the three
most common averages and exactly when each should be applied.
We will end by comparing the averages when they are applied to the
same numeric values.

The Arithmetic Average

THE MOST COMMON average of all is the one you probably learned
in late elementary school, or in middle school, but almost certainly

5.6. TYPES OF AVERAGES

before high school. When values are summable, as most things are, this is the average which should be used. Distance, time, weight, age, prices of individual items, counts are all summable quantities. Any quantity that represents something physical is summable. It is so commonly used that when people hear the words "take the average of n values" this is the average which comes to mind. In fact, it may be the *only* average you ever learned.

The **arithmetic average**, denoted $\overline{x}$, is the sum of the quantities being averaged divided by their number; that is, if $x_1, x_2, \ldots, x_n$ are the values being averaged then

$$\overline{x} = \frac{\sum_{i=1}^{n} x_i}{n}.$$

When some values are repeated the formula looks a little different. In reality it is exactly the same. Suppose there are k distinct values $y_1, y_2, \ldots, y_k$ and that $\text{frq}(y_i) = f_i$. Then

$$\overline{y} = \frac{\sum_{i=1}^{k} y_i \cdot f_i}{\sum_{i=1}^{k} f_i}.$$

Example 1:
Determine the arithmetic average for the oyster data.

Answer:
Forty-seven of the fifty data are unique. It is not worth the effort to use the frequency form. The sorted data are repeated here:

5.76	6.10	6.36	6.37	6.42	6.53	6.63	6.67	6.74	6.91
6.91	6.97	7.03	7.04	7.37	7.40	7.71	7.97	8.02	8.04
8.12	8.14	8.42	8.53	8.61	8.62	9.01	9.20	9.21	9.39
9.47	9.53	9.55	9.98	10.24	10.50	10.62	10.63	10.63	10.81
10.81	10.86	11.02	11.09	11.17	11.31	11.59	11.97	12.07	13.18

This makes

$$\bar{x} = \frac{5.76 + 6.10 + \cdots + 13.18}{50} = \frac{443.23}{50} = \frac{44323}{5000} = 8.8646 \approx 8.86$$

The arithmetic average answers: *What single value would I need to replace each quantity by to get the same sum?* If you cannot add up the items, then there cannot be a sum; whence, the need for the items to be summable.

Example 2:
There are eight stores in your area that sell a particular item. At three stores, the price of this item was $26.00. At three other stores, the price was $30.00. Finally, at the other two stores the price of the item is $32.00. What is the average selling price for this item in your area?

Answer:
If we assume that each store has the same chance to sell the item as every other store, then the arithmetic average applies. There are three prices: $p_1 = 26$ with $f_1 = 3$; $p_2 = 30$ with $f_2 = 3$; and $p_3 = 32$ with $f_3 = 2$. Whereby

$$\bar{p} = \frac{26 \cdot 3 + 30 \cdot 3 + 32 \cdot 2}{3 + 3 + 2} = \frac{78 + 90 + 64}{8} = 29.$$

We conclude the average selling price is $29.00.

5.6. TYPES OF AVERAGES

Nowhere can you actually purchase the item for $29.00, but if you bought one item from each store you would pay $232.00 for the eight items. Thus, on average, you spent $29.00 per item as $8 \cdot 29 = 232$.

The Geometric Average

QUANTITIES WHICH ARE not summable are **rates**. There are two kinds of rates. Here we consider factorisable rates; later we will discuss independent rates. A **factorisable rate** is a *periodic* rate of change. Thus, these rates have a time component about them. The final amount depends upon the product of these rates, one per period. The **geometric average** is the average rate of change per period. Rates of decay, rates of growth, and yield on an investment are all factorisable rates. Slope is not.

The **geometric average**, denoted $\tilde{x}$, is the product of the quantities being averaged taken to the root of their number; that is, if x_1, x_2, ..., x_n are the values being averaged then

$$\tilde{x} = \sqrt[n]{\prod_{i=1}^{n} x_i}.$$

When some values are repeated the formula looks a little different, in reality it is exactly the same. Suppose there are k distinct values y_1, y_2, ..., y_k and that $\text{frq}(y_i) = f_i$. Then

$$\tilde{y} = \sqrt[n]{\prod_{i=1}^{k} y_i^{f_i}},$$

where

$$n = \sum_{i=1}^{k} f_i.$$

Example 3:
You wish to invest in a money market fund. Te money market fund has had the following percentage returns for the last ten years:

$$5 \quad 2 \quad 4 \quad \text{-}2 \quad \text{-}2 \quad 2 \quad 4 \quad 4 \quad 2 \quad 2$$

What is the average return for this money market?

Answer:
The period rate of change is **not** r, it is $1 + r$. The r values (as percentages) are $r_1 = \text{-}2$, $r_2 = 2$, $r_3 = 4$ and $r_4 = 5$. Which makes $x_1 = 0.98$, $x_2 = 1.02$, $x_3 = 1.04$ and $x_4 = 1.05$. Their frequencies are $f_1 = 2$, $f_2 = 4$, $f_3 = 4$ and $f_4 = 1$. Observe $n = 2 + 4 + 3 + 1 = 10$ — the ten years.

$$\tilde{x} = \sqrt[n]{\prod_{i=1}^{4} x_i^{f_i}} = \sqrt[10]{0.98^2 \cdot 1.02^4 \cdot 1.04^3 \cdot 1.05^1}$$
$$= \sqrt[10]{1.2278410683471249408 0} \approx 1.02073784.$$

The average annual rate of return is $\tilde{r} \approx 2.074\%$.

The geometric average answers the question: *What single value would I need to replace each quantity by to get the same product?* If you cannot multiply the items, then there cannot be a product; whence, the need for the rates to be factorisable.

Example 4:
The average inflation rate over the last five years was reported as 3%. If prices went up last year by 5% then what was the average inflation rate for the previous four years?

Answer:

5.6. TYPES OF AVERAGES

The overall product is
$$1.03^5 = 1.1592740743.$$
We are told that last year the rate was 1.05; whence,
$$1.05 \cdot \tilde{x}^4 = 1.1592740743.$$
We conclude
$$\tilde{x} = \sqrt[4]{\frac{1.1592740743}{1.05}} \approx 1.0250598.$$
The average inflation rate for the previous four years was 2.506%.

Example 5:
What rate, compounded quarterly, is equivalent to an annual rate of 5%?

Answer:
The overall product is 1.05 and there are 4 factors; thereby,
$$\tilde{x} = \sqrt[4]{1.05} \approx 1.0122722.$$
The periodic rate is 1.22722%. This is usually reported as $4 \cdot 1.22722 \approx 4.909\%$, compounded quarterly.

The Harmonic Average

THERE ARE TWO kinds of rates. We discussed factorisable rates earlier; here we discuss independent rates. An **independent rate** is the ratio of two summable quantities. They obey the general formula
$$\text{rate} \cdot \text{quantity} = \text{frequency}.$$

Slope (with quantity the change in x and frequency the change in y), speed (with quantity time and frequency distance), share price (with quantity the number of shares and frequency the amount invested) and resistance of a parallel circuit (with quantity the current and frequency number of resistors) are all independent rates.

The **harmonic average**, denoted $\check{x}$, is the sum of the frequencies divided by the sum of the quantities. If $x_1, x_2, \ldots, x_n$ are the independent rates being averaged, then

$$\check{x} = \frac{n}{\sum_{i=1}^{n} \frac{1}{x_i}}.$$

When some values are repeated the formula looks a little different, in reality it is exactly the same. Suppose there are k distinct values $y_1, y_2, \ldots, y_k$ and that $\text{frq}(y_i) = f_i$. Then from the defining formula,

$$y_i \cdot q_i = f_i; \text{ so, } q_i = \frac{f_i}{y_i}.$$

This means

$$\check{y} = \frac{\sum_{i=1}^{k} f_i}{\sum_{i=1}^{k} q_i} = \frac{\sum_{i=1}^{k} f_i}{\sum_{i=1}^{k} \frac{f_i}{y_i}}.$$

Example 6:
You live 30 kilometers from where you work. You leave early, so morning traffic is not a problem and you can average 90 kph without any trouble. When you come home, however, it is rush hour and you

5.6. TYPES OF AVERAGES

can only average 50 kph on the way home. What is your average speed in both directions?

Answer:
The distance traveled is the frequency, so $x_1 = 90$ and $f_1 = 30$ and $x_2 = 50$ and $f_2 = 30$. We conclude

$$\check{x} = \frac{30 + 30}{\frac{30}{90} + \frac{30}{50}} = \frac{60}{14/15} = \frac{450}{7} \approx 64.286 \text{ kph}.$$

Example 7:
You invested $1000 to buy into a money market account when the share price was $250. You later invested $1125 when the share price went to $225, $1000 when it went back down to $200, and $1500 when the share price went to $250. What is your average cost per share?

Answer:
The amount invested is the frequency. The first and last share prices are the same hence $x_1 = 250$ with $f_1 = 1000 + 1500 = 2500$, $x_2 = 225$ with $f_2 = 1125$ and $x_3 = 200$ with $f_3 = 1000$. We conclude

$$\check{x} = \frac{2500 + 1125 + 1000}{\frac{2500}{250} + \frac{1125}{225} + \frac{1000}{200}} = \frac{4625}{20} = \frac{925}{4} = \$231.25.$$

Example 8:
One 10 Ω and two 5 Ω resistors are connected in parallel. Determine the equivalent resistance for this circuit.

Answer:

The equivalent resistance of resistors connected in parallel is the harmonic average of the the individual resistors: the frequency is the number of resistors and the current passing through each resistor is the quantity. Thus,
$$\check{x} = \frac{1+2}{\dfrac{1}{10}+\dfrac{2}{5}} = \frac{3}{1/2} = 6\,\Omega.$$

Comparing Averages

THERE ARE NO SIGN restrictions for the arithmetic average. This is not true for the other averages.

If the number of items is odd then there are no sign restrictions in the geometric average; however, if the number of items is *even* then the product must be nonnegative. We conclude that it is best to restrict values to being nonnegative when a geometric average is needed.

Zero is not allowed in the harmonic average, nor is a sign change, thus all items must be either positive or all must be negative. It is best to restrict things to positive numbers as these are more commonly encountered.

The above restrictions just mean that if all observations are positive then all three types of averages could be taken. [We will not discuss why you might want to do this as it would make no sense in a practical problem.]

One more thing: the frequencies in the formulas for the three averages do *not* have to be integers — although in practice it is rare to see them not *be* integers.

5.6. TYPES OF AVERAGES

Theorem 5.1 (Comparison of Averages). *Suppose $x_1, x_2, \ldots, x_k$ are positive real numbers and that $\mathrm{frq}(x_i) = f_i$. Then*

$$\check{x} \leq \tilde{x} \leq \bar{x};$$

that is, the harmonic average is smallest and the arithmetic average is largest among the three averages.

Proof. Omitted. □

Example 9:
Verify Theorem 5.1 holds for the values below.

$$2 \quad 4 \quad 2 \quad 3 \quad 2 \quad 3 \quad 5 \quad 2 \quad 4 \quad 3$$

Answer:
There are 10 values: $x_1 = 2$ with $f_1 = 4$, $x_2 = 3$ with $f_2 = 3$, $x_3 = 4$ with $f_3 = 2$ and $x_4 = 5$ with $f_4 = 1$.
The harmonic average is

$$\check{x} = \frac{4+3+2+1}{\frac{4}{2}+\frac{3}{3}+\frac{2}{4}+\frac{1}{5}} = \frac{10}{37/10} = \frac{100}{37} \approx 2.7027.$$

The geometric average is (Note: $4+3+2+1 = 10$.)

$$\tilde{x} = \sqrt[10]{2^4 \cdot 3^3 \cdot 4^2 \cdot 5^1} = \sqrt[10]{16 \cdot 27 \cdot 16 \cdot 5} = \sqrt[10]{34560} \approx 2.8435.$$

The arithmetic average is

$$\bar{x} = \frac{2 \cdot 4 + 3 \cdot 3 + 4 \cdot 2 + 5 \cdot 1}{4+3+2+1} = \frac{8+9+8+5}{10} = 3.$$

Observe:
$$2.7027 \leq 2.8435 \leq 3$$

as Theorem 5.1 claimed it would.

Exercises 5.6

1. Determine the arithmetic average of the following quantities:

   16 93 42 42 51 25 23 93 91 57

2. Determine the arithmetic average of the following quantities:

   0.96 1.08 1.03 1.05 1.08 0.96 1.03 1.08 1.03 0.96

3. Determine the geometric average of the following quantities:

   16 93 42 42 51 25 23 93 91 57

4. Determine the geometric average of the following quantities:

   0.96 1.08 1.03 1.05 1.08 0.96 1.03 1.08 1.03 0.96

5. Determine the harmonic average of the following quantities:

   16 93 42 42 51 25 23 93 91 57

6. Determine the harmonic average of the following quantities:

   0.96 1.08 1.03 1.05 1.08 0.96 1.03 1.08 1.03 0.96

5.7 Mean

THE ARITHMETIC average of the data is the most commonly used statistic for the most representative value. The **mean**, denoted $\bar{x}$, is the name for this statistic. Suppose each observed value is placed along a number line and given a mass equal to its frequency in kilograms. If a fulcrum were placed at the point equal to the mean, then the number line would perfectly balance. For this reason it is

5.7. MEAN

considered the most representative value. As the arithmetic average, the formula for the mean is

$$\bar{x} = \frac{\sum_{i=1}^{n} x_i}{n}.$$

In terms of the distinct observed values, the formula looks slightly different. It is, of course, exactly the same. Suppose $y_1, y_2, \ldots, y_k$ distinct observed values and that $\text{frq}(y_i) = f_i$. Then

$$\bar{y} = \frac{\sum_{i=1}^{k} y_i \cdot f_i}{\sum_{i=1}^{n} f_i}.$$

If the population is finite then we can also talk about the **population mean**, denoted μ. Its formulas are identical to those above, except now n is the size of the entire population and there are only k distinct values in the entire population. In general, including infinite populations, the population mean is also often called the **expected value**. It is not expected in the normal sense, as actually expecting to see it, but rather in the sense that if you repeated choose samples from the population and compute their mean, then these averages are all close to it and the average of all these averages approaches it. We will talk more about this later with the restriction that all of the samples have the same number of observations.

Example 1:
Calculate the mean of the two-success data.

Answer:
The frequency table is repeated below.

```
2:  9    6: 9    10: 6    15: 1
3: 16    7: 4    11: 4    16: 1
4: 15    8: 4    12: 2    17: 1
5: 17    9: 5    14: 5    28: 1
```

From the above
$$\bar{x} = \frac{2 \cdot 9 + 3 \cdot 16 + \cdots + 28 \cdot 1}{9 + 16 + \cdots + 1} = \frac{644}{100} = 6.44.$$

Example 2:
Determine the mean for the random gamma data.

Answer:
The frequency table is repeated below.

```
4: 1     9: 2    13: 3    17: 2    24: 2
5: 4    10: 1    14: 2    18: 1    27: 1
8: 1    12: 1    15: 1    22: 2    31: 1
```

Using the data above,
$$\bar{x} = \frac{4 \cdot 1 + 5 \cdot 4 + \cdots + 31 \cdot 1}{1 + 4 + \cdots + 1} = \frac{356}{25} = 14.24.$$

Example 3:
Find the mean of the dice roll data.

Answer:
The sorted data are repeated below.

```
2  3  3  3  3  3  4  4  4  4
5  5  5  5  5  5  6  6  6  6
6  6  6  7  7  7  7  7  7  7
7  8  8  8  8  8  8  8  9  9
9  9  9  10 10 10 10 10 11 12
```

5.7. MEAN

Therefrom,
$$\bar{x} = \frac{2+3+3+\cdots+12}{50} = \frac{335}{50} = 6.7.$$

Grouped Data

SOMETIMES ONE only has access to grouped data and not the actual measurements. It is still possible to calculate a mean from such data. As we did for Walsh sums, we use the midpoints of each interval as the observation and the number of observations which fell in that interval as the frequency for its midpoint.

Example 4:
Determine the mean for the two success data when the data are divided originally into six intervals. Then merge the empty interval into the interval following, the interval preceding, and split it proportionately among the two touching intervals and in each case recalculate the mean.

Answer:
When there are six intervals, the common width is
$$w = \frac{28-2}{6} = \frac{13}{3} \approx 4.333$$

We round this to the nearest 0.1 and make $w = 4.4$. For the purpose of determining endpoints for the intervals we set
$$t_0 = \frac{2 + 28 - 6 \cdot 4.4}{2} = 1.8,$$

but realize that it must start at $\mu_{\min} = 2$. This gives the following intervals and their frequencies.

I_i	$[2, 6.2)$	$[6.2, 10.6)$	$[10.6, 15)$	$[15, 19.4)$	$[19.4, 23.8)$	$[23.8, 28.2]$
m_i	4.1	8.4	12.8	17.2	21.6	26
f_i	66	19	11	3	0	1

We conclude from the above that

$$\bar{x} = \frac{4.1 \cdot 66 + 8.4 \cdot 19 + \cdots + 26 \cdot 1}{66 + 19 + \cdots + 1} = \frac{648.6}{100} = 6.486.$$

Combining the last two intervals creates $[19.4, 28.2]$ which has midpoint 23.8 and frequency 1. The new estimate is

$$\bar{x} = \frac{4.1 \cdot 66 + 8.4 \cdot 19 + \cdots + 23.8 \cdot 1}{66 + 19 + \cdots + 1} = \frac{646.4}{100} = 6.464.$$

Combining the empty interval with the previous interval makes an interval $[15, 23.8)$ which has midpoint 19.4 and frequency 3. The revised estimate is

$$\bar{x} = \frac{4.1 \cdot 66 + 8.4 \cdot 19 + \cdots + 26 \cdot 1}{66 + 19 + \cdots + 1} = \frac{655.2}{100} = 6.552.$$

The sum of the frequencies for the touching intervals is 4. If we take 3/4 of the interval for the previous and 1/4 for the next interval then the three intervals become two. The interval $[15, 22.7)$ has midpoint 18.85 and frequency 3. The interval $[22.7, 28.2]$ has midpoint 25.45 and frequency 1. Our final estimate is

$$\bar{x} = \frac{4.1 \cdot 66 + 8.4 \cdot 19 + \cdots + 25.45 \cdot 1}{66 + 19 + \cdots + 1} = \frac{653}{100} = 6.53.$$

Exercises 5.7

1. One day while walking through a field you notice there seem to be a lot of stones lying on the ground. Curious about what the average

5.7. MEAN

size of a stone is in the field you randomly collect 30 stones. You weigh each stone and record its weight in grams. The following table shows the raw data.

110	126	75	110	126	94
90	112	111	139	131	92
102	103	96	75	67	101
104	111	100	65	92	93
106	104	93	115	109	95

Determine the mean of the data. Then group the data into five intervals and calculate the mean for the grouped data.

2. The scores, in the order graded, on a recent midterm examination in Statistics, a class with 30 students, are given below.

70	80	99	98	85	89
87	79	83	38	69	70
60	69	78	40	75	56
70	51	99	69	95	86
57	53	47	50	55	81

Determine the mean of the data. Then group the data using one of the grading scales to determine the intervals and determine the mean again. The grading scales are

American		International	
At Least	Grade	At Least	Grade
90	A	90	Excellent
80	B	75	Good
70	C	60	Acceptable
60	D	50	Poor
0	F	0	Fail

3. Below is a sample of the monthly salary for 30 randomly selected people in mid-management in thousands of dollars.

3.08	3.18	3.13	7.41	3.73	3.05
4.17	3.58	3.36	3.27	3.36	4.74
3.32	3.61	4.26	3.02	3.45	4.06
6.73	4.72	3.13	3.15	3.70	3.59
3.12	3.61	5.03	3.53	3.20	3.32

Determine the mean of the data. Then group the data into five intervals and calculate the mean for the grouped data.

4. A pair of dice, one red and one green, is rolled in the following manner. On the first toss both dice are tossed. Then the red die is tossed while the green die keeps its value to get the second sum. Then the green die is tossed while the red die keeps its value. You continue tossing alternately the red then the green, die and generate the following table.

8	11	8	6	10	9	9
7	2	5	10	11	8	5
3	6	9	7	7	9	10
7	3	3	7	9	8	7
5	4	8	11	11	10	10

Determine the mean of the data. Then group the data into five intervals and calculate the mean for the grouped data.

5. The following random sample of 72 observations comes from a gamma distribution, so $\mu_{min} = 0$ and μ_{max} has no upper limit. The raw data are

2.15	1.92	1.92	2.20	3.67	4.55	3.32	2.71	4.87
2.09	1.13	1.09	2.12	4.89	3.03	3.17	4.74	4.44
4.61	2.78	2.97	2.78	2.88	3.66	2.55	4.40	3.54
2.18	3.43	3.84	4.60	1.76	2.85	2.27	1.38	1.24
4.22	1.67	4.51	3.50	6.69	5.92	6.42	0.95	4.57
3.10	3.26	2.95	2.45	1.95	2.06	3.10	2.03	4.98
0.86	2.61	1.92	2.78	1.01	2.64	3.14	0.86	3.68
2.71	3.06	4.33	4.63	2.47	2.92	1.38	2.81	5.71

Determine the mean of the data. Then group the data into five, six and seven intervals and calculate the mean for the grouped data.

5.8 Tukey's Trimean

THE PROBLEM with the mean is that it is strongly influenced by outliers. This is one reason why many prefer the median. Another statistic which is not influenced by outliers is Tukey's trimean. It should probably be called Bowley's trimean, as he is the one who proposed it, but it is called Tukey's trimean because it first appeared in Tukey's book on exploratory data analysis in 1978.[1]

Tukey's trimean is a weighted average of the quartiles. We will use $\mathfrak{M}$ for this statistic. Specifically,

$$\mathfrak{M} = \frac{Q_1 + 2M + Q_3}{4}.$$

The quartiles are normally computed using the traditional sample method or the restricted proper method. However, the median-of-medians method is also widely used. The statistic can also be calculated for grouped data using the grouped method to calculate the quartiles.

Example 1:
Determine Tukey's trimean for the random gamma data.

Answer:
    In Example 1, §4.6, which started on page 165, we determined the quartiles using a variety of methods. The results are summarized here.

---

[1] This happens from time to time. Cramer's rule in linear algebra got its name for a similar reason but was created by one of the Bernoullis. Cramer spent the next 30 years trying to get people to call it Bernoulli's rule, but no one listened. It is still called Cramer's rule.

Method	Q_1	M	Q_3
Averaging			
Nearest Averaging	8.50	13.00	20.00
Estimation			
Nearest or Even	8.00	13.00	22.00
Median-of-Medians	9.00	13.00	17.00
Regular Proper	8.33	13.40	19.33
Restricted Proper	8.67	13.40	18.00
Traditional Sample	9.00	13.00	18.00
Four Intervals	8.36	13.22	18.55
Five Intervals	8.22	12.87	19.86

We will calculate Tukey's trimean in each case.

For the averaging, nearest and estimation methods
$$\mathfrak{M} = \frac{8.5 + 2 \cdot 13 + 20}{4} = 13.625 \approx 13.62.$$

For the nearest method which uses the even observation when encountering a half-integer, we get
$$\mathfrak{M} = \frac{8 + 2 \cdot 13 + 22}{4} = 14.$$

For the median-of-medians method, Tukey's trimean is
$$\mathfrak{M} = \frac{9 + 2 \cdot 13 + 17}{4} = 13.$$

For the regular proper method, using the rounded values from the table, we get
$$\mathfrak{M} = \frac{8.33 + 2 \cdot 13.4 + 19.33}{4} = 13.615 \approx 13.62.$$

For the restricted proper method, again using the rounded values, we obtain
$$\mathfrak{M} = \frac{8.67 + 2 \cdot 13.4 + 18}{4} = 13.3675 \approx 13.37.$$

Tukey's trimean using the traditional sample method is
$$\mathfrak{M} = \frac{9 + 2 \cdot 13 + 18}{4} = 13.25.$$

5.8. TUKEY'S TRIMEAN

Using four intervals and the grouped method makes Tukey's trimean, from the rounded entries,

$$\mathfrak{M} = \frac{8.36 + 2 \cdot 13.22 + 18.55}{4} = 13.3375 \approx 13.34.$$

Finally, the grouped method with five intervals, and using the rounded entries, makes Tukey's trimean

$$\mathfrak{M} = \frac{8.22 + 2 \cdot 12.87 + 19.86}{4} = 13.455 \approx 13.46.$$

Example 2:
Calculate Tukey's trimean for the two-success data using each of the following methods: median-of-medians, restricted proper and traditional sample. Then group the data into six intervals — do not worry about the empty interval — and calculate the statistic again.

Answer:
The tally table for this data is

2: 9	6: 9	10: 6	15: 1
3: 16	7: 4	11: 4	16: 1
4: 15	8: 4	12: 2	17: 1
5: 17	9: 5	14: 5	28: 1

There are 100 data values.
In the median-of-medians method, $\#(M) = (1+100)/2 = 50.5$. Now, $x_{(50)} = x_{(51)} = 5$, so $M = 5$. There are 57 values which are less than or equal to 5, thus,

$$\#(Q_1) = \frac{1+57}{2} = 29 \quad \text{and} \quad Q_1 = x_{(29)} = 4.$$

There are 60 values greater than or equal to 5, the first is $x_{(41)} = 5$ and the last is $x_{(100)} = 28$. Therefore,

$$\#(Q_3) = \frac{41+100}{2} = 70.5.$$

Now, $x_{(70)} = 7$ and $x_{(71)} = 8$, so

$$Q_3 = \frac{7+8}{2} = 7.5.$$

Using the above information,

$$\mathfrak{M} = \frac{4 + 2 \cdot 5 + 7.5}{4} = 5.375.$$

For the restricted proper method,

$$\#(2) = \frac{1+9}{2} = 5 \quad \text{and} \quad \#(28) = 100.$$

This means

$$\#(Q_1) = \frac{3 \cdot 5 + 100}{4} = 28.75$$
$$\#(M) = \frac{5 + 100}{2} = 52.5$$
$$\#(Q_3) = \frac{5 + 3 \cdot 100}{4} = 76.25.$$

For Q_1, $x_{(28)} = 4$ and $\#(4) = (26 + 40)/2 = 33 > 28.75$. Therefore, $y_2 = 4$, $m_2 = 33$, $y_1 = 3$ and $m_1 = (10 + 25)/2 = 17.5$. By interpolation,

$$Q_1 = \frac{3 \cdot (33 - 28.75) + 4 \cdot (28.75 - 17.5)}{33 - 17.5}$$
$$= \frac{12.75 + 45}{15.5} = \frac{231}{62} \approx 3.72581.$$

For M, $x_{(52)} = 5$ and $\#(5) = (41 + 57)/2 = 49 < 52.5$. Whereby, $y_1 = 5$, $m_1 = 49$, $y_2 = 6$ and $m_2 = (58 + 66)/2 = 62$. By interpolation,

$$M = \frac{5 \cdot (62 - 52.5) + 6 \cdot (52.5 - 49)}{62 - 49} = \frac{47.5 + 21}{13} = \frac{137}{26} \approx 5.26923.$$

5.8. TUKEY'S TRIMEAN

For Q_3, $x_{(76)} = 9$ and $\#(9) = (75+79)/2 = 77 > 76.25$. Thereby, $y_2 = 9$, $m_2 = 77$, $y_1 = 8$ and $m_1 = (71+74)/2 = 72.5$. By interpolation,

$$Q_3 = \frac{8 \cdot (77 - 76.25) + 9 \cdot (76.25 - 72.5)}{77 - 72.5}$$
$$= \frac{6 + 33.75}{4.5} = \frac{53}{6} \approx 8.83333.$$

From the above rounded values,

$$\mathfrak{M} = \frac{3.72581 + 2 \cdot 5.26923 + 8.83333}{4} = 5.7744.$$

In the traditional sample method,

$$\#(Q_1) = \frac{100}{4} = 25 \quad \#(M) = \frac{100}{2} = 50 \quad \#(Q_3) = 3\#(Q_1) = 75.$$

For Q_1,

$$Q_1 = \frac{x_{(25)} + x_{(26)}}{2} = \frac{3+4}{2} = 3.5.$$

For M, $M = 5$ as $x_{(50)} = x_{(51)} = 5$. For Q_3, $Q_3 = 9$ as $x_{(75)} = x_{(76)} = 9$. Therefore, Tukey's trimean is

$$\mathfrak{M} = \frac{3.5 + 2 \cdot 5 + 9}{4} = 5.625.$$

The six intervals and their frequencies were calculated in Example 4, §5.7, starting on page 231. These are

I_i	[2, 6.2)	[6.2, 10.6)	[10.6, 15)	[15, 19.4)	[19.4, 23.8)	[23.8, 28.2]
m_i	4.1	8.4	12.8	17.2	21.6	26
f_i	66	19	11	3	0	1

The effective positions of the quartiles are the same as in the traditional sample method, so

$$\#(Q_1) = 25 \quad \#(M) = 50 \quad \#(Q_3) = 75.$$

The first two are in the first interval and the third is in the second interval — this is why we do not worry about the empty interval; it has no effect on the results.

For Q_1,
$$Q_1 = 2 + \frac{25 \cdot (6.2 - 2)}{66} = \frac{79}{22} \approx 3.59091.$$

For M,
$$M = 2 + \frac{50 \cdot (6.2 - 2)}{66} = \frac{57}{11} \approx 5.18182.$$

For Q_3,
$$\text{offset} = 75 - 66 = 9 \leq 19.$$

Therefore,
$$Q_3 = 6.2 + \frac{9 \cdot (10.6 - 6.2)}{19} = \frac{787}{95} \approx 8.28421.$$

Using the above rounded values,
$$\mathfrak{M} = \frac{3.59091 + 2 \cdot 5.18182 + 8.28421}{4} = 5.55969 \approx 5.560.$$

Exercises 5.8

1. One day while walking through a field you notice there seem to be a lot of stones lying on the ground. Curious about what the average size of a stone is in the field you randomly collect 30 stones. You weigh each stone and record its weight in grams. The following table shows the raw data.

110	126	75	110	126	94
90	112	111	139	131	92
102	103	96	75	67	101
104	111	100	65	92	93
106	104	93	115	109	95

Determine Tukey's trimean for the data. Then group the data into five intervals and calculate Tukey's trimean for the grouped data.

2. The scores, in the order graded, on a recent midterm examination in Statistics, a class with 30 students, are given below.

70	80	99	98	85	89
87	79	83	38	69	70
60	69	78	40	75	56
70	51	99	69	95	86
57	53	47	50	55	81

Determine Tukey's trimean for the data. Then group the data using one of the grading scales to determine the intervals and determine Tukey's trimean again. The grading scales are

American		International	
At Least	Grade	At Least	Grade
90	A	90	Excellent
80	B	75	Good
70	C	60	Acceptable
60	D	50	Poor
0	F	0	Fail

3. Below is a sample of the monthly salary for 30 randomly selected people in mid-management in thousands of dollars.

3.08	3.18	3.13	7.41	3.73	3.05
4.17	3.58	3.36	3.27	3.36	4.74
3.32	3.61	4.26	3.02	3.45	4.06
6.73	4.72	3.13	3.15	3.70	3.59
3.12	3.61	5.03	3.53	3.20	3.32

Determine Tukey's trimean for the data. Then group the data into five intervals and calculate Tukey's trimean for the grouped data.

4. A pair of dice, one red and one green, is rolled in the following manner. On the first toss both dice are tossed. Then the red die is tossed while the green die keeps its value to get the second sum. Then the green die is tossed while the red die keeps its value. You continue tossing alternately the red then the green, die and generate the following table.

```
8  11  8   6  10   9   9
7   2  5  10  11   8   5
3   6  9   7   7   9  10
7   3  3   7   9   8   7
5   4  8  11  11  10  10
```

Determine Tukey's trimean for the data. Then group the data into five intervals and calculate Tukey's trimean for the grouped data.

5. The following random sample of 72 observations comes from a gamma distribution, so $\mu_{min} = 0$ and μ_{max} has no upper limit. The raw data are

```
2.15  1.92  1.92  2.20  3.67  4.55  3.32  2.71  4.87
2.09  1.13  1.09  2.12  4.89  3.03  3.17  4.74  4.44
4.61  2.78  2.97  2.78  2.88  3.66  2.55  4.40  3.54
2.18  3.43  3.84  4.60  1.76  2.85  2.27  1.38  1.24
4.22  1.67  4.51  3.50  6.69  5.92  6.42  0.95  4.57
3.10  3.26  2.95  2.45  1.95  2.06  3.10  2.03  4.98
0.86  2.61  1.92  2.78  1.01  2.64  3.14  0.86  3.68
2.71  3.06  4.33  4.63  2.47  2.92  1.38  2.81  5.71
```

Determine Tukey's trimean for the data. Then group the data into five, six and seven intervals and calculate Tukey's trimean for the grouped data.

5.9 Trimmed Means

TUKEY'S TRIMEAN is based upon the **core**, the central fifty percent of the data. As such it is not influenced by outliers. There are two objections people have with it. First, you are ignoring the contribution outside of the core. Second, you are not utilizing the full power of the data even within the core, only the edges and "middle." They

5.9. TRIMMED MEANS

reason that if the mean is being influenced by outliers, then the outliers are going to be rare, so you really only need to "trim the ends up a bit."

You cannot trim off only only the outliers because there might only be outliers on one end of the data and you would be imposing a bias toward the side you keep which could unbalance the rest of the data and you end with a vicious cycle of trimming. There are two ways to perform the trimming: You can trim off a fixed number from each end; or, you can trim off a fixed percentage from each end. We will consider both methods of trimming.

r-Trimmed Mean

PROBABLY THE easiest way of performing the trimming is to remove a fixed number from each end. The **r-trimmed mean** removes r observations from each end, $2r$ observations altogether. It is denoted $\overline{T_r}$. and has the formula

$$\overline{T_r} = \frac{\sum_{i=r+1}^{n-r} x_{(i)}}{n - 2 \cdot r}.$$

You can take advantage of frequency in calculating, but the general form is a nightmare and will not be shown — although we *will* demonstrate it.

Example 1:
Calculate the quartiles in the two-success data using the median-of-medians, traditional sample, and restricted proper methods. Use the three-halves rule to calculate the step size in each case and determine

the number of outliers on each end. Make r the maximum number of outliers on an end and calculate the r-trimmed mean for the data.

Answer:

The tally table for the data is

2: 9	4: 15	6: 9	8: 4	10: 6	12: 2	15: 1	17: 1
3: 16	5: 17	7: 4	9: 5	11: 4	14: 5	16: 1	28: 1

The quartiles for each of these methods was determined in Example 2, §5.8, which started on page 237. These are summarized as

Method	Q_1	M	Q_3
Median-of-Medians	4.00000	5.00000	7.50000
Traditional Sample	3.50000	5.00000	9.00000
Restricted Proper	3.72581	5.26923	8.83333

By the median-of-medians method, the interquartile range is $R_Q = 7.5 - 4 = 3.5$, so the step size is $1.5 \cdot 3.5 = 5.25$. The fence is

$$F_L = \max(4 - 5.25, 2) = 2 \quad F_U = \min(7.5 + 5.25, 28) = 12.75.$$

This makes 9 outliers: 14×5, 15, 16, 17 and 28, all on the topside.

$$\bar{T}_9 = \frac{\sum_{i=10}^{91} x_{(i)}}{100 - 2 \cdot 9} = \frac{3 \cdot 16 + \cdots + 12 \cdot 2}{82} = \frac{240}{41} \approx 5.854.$$

For the traditional sample method, the interquartile range is $R_Q = 9 - 3.5 = 5.5$, so the step size is $1.5 \cdot 5.5 = 8.25$. The fence is

$$F_L = \max(3.5 - 8.25, 2) = 2 \quad F_U = \min(9 + 8.25, 28) = 17.25.$$

From this, only 28 is an outlier.

$$\bar{T}_1 = \frac{\sum_{i=2}^{99} x_{(i)}}{100 - 2 \cdot 1} = \frac{2 \cdot 8 + 3 \cdot 16 + \cdots + 17 \cdot 1}{98} = \frac{307}{49} \approx 6.265.$$

5.9. TRIMMED MEANS

For the restricted proper method, the interquartile range is $R_Q = 8.83333 - 3.72581 = 5.10752$, so the step size is $1.5 \cdot 5.10752 = 7.66128$. The fence is
$$F_L = \max(3.72581 - 7.66128, 2) = 2$$
and
$$F_U = \min(8.83333 + 7.66128, 28) = 16.49461.$$
This makes 17 and 28 outliers — both on the topside.

$$T_2 = \frac{\sum_{i=3}^{98} x_{(i)}}{100 - 2 \cdot 2} = \frac{2 \cdot 7 + 3 \cdot 16 + \cdots + 16 \cdot 1}{96} = \frac{595}{96} \approx 6.198.$$

Note: What made the median-of-medians method so restrictive is that 17 values equal the median — more that one-sixth of the data. This made the quartiles exceptionally close together

We could not have used the averaging, nearest or estimation methods because $\mu_{\min} = 2$ is an observed value. Traditionally, none of these is used anyway. It would be possible to use the regular proper method, but we leave that to you to show that the outliers are the same as for the restricted proper method when the three-halves rule is used. You might want to see how things change when the three-eighths rule is used.

Example 2:
Determine whether there are any outliers in the random gamma data using the median-of-medians method. If so, use the appropriate r-trimmed mean to eliminate them.

Answer:

In Example 3, §4.6, which started on page 173, we found that the median-of-medians method determined that $x = 31$ is the only outlier. Therefore, we need to calculate the 1-trimmed mean, T_1, for this data. The frequency table is repeated below.

4: 1	9: 2	13: 3	17: 2	24: 2
5: 4	10: 1	14: 2	18: 1	27: 1
8: 1	12: 1	15: 1	22: 2	31: 1

From the above data, dropping off the first and last sorted observations, we get

$$\overline{T}_1 = \frac{\sum_{i=2}^{24} x_{(i)}}{25 - 2 \cdot 1} = \frac{5 \cdot 4 + \cdots + 27 \cdot 1}{23} = \frac{321}{23} = 13.957.$$

The r-trimmed mean is related to the traditional sample median. For $r = \lfloor (n-1)/2 \rfloor$, $M = \overline{T}_r$.

p-Trimmed Mean

GIVEN A PERCENTILE rank p, we calculate indices $s = \lfloor \#q(p) \rfloor + 1$ and $t = \lfloor \#q(1-p) \rfloor$. The **$p$-trimmed mean**, denoted $\boldsymbol{T(p)}$, is

$$\overline{T}(p) = \frac{\sum_{i=s}^{t} x_{(i)}}{t - s + 1}.$$

Example 3:
Determine the 10%-trimmed mean for the two-success data.

Answer:
The tally table for the data is

5.9. TRIMMED MEANS

2: 9	4: 15	6: 9	8: 4	10: 6	12: 2	15: 1	17: 1
3: 16	5: 17	7: 4	9: 5	11: 4	14: 5	16: 1	28: 1

There are 100 observations in the sample.
By the traditional method,

$$\#q(0.1) = 0.1 \cdot 100 = 10 \text{ and } s = 11;$$

Furthermore,

$$\#q(0.9) = 0.9 \cdot 100 = 90 \text{ and } t = 90;$$

we conclude

$$\bar{T}(0.1) = \frac{\sum_{i=11}^{90} x_{(i)}}{90 - 11 + 1} = \frac{3 \cdot 15 + \cdots + 12 \cdot 1}{80} = \frac{93}{16} = 5.8125.$$

By the restricted proper method

$$\#(x_{\min}) = \frac{1+9}{2} = 5 \text{ and } \#(x_{\max}) = 100.$$

Thereby,

$$\#q(0.1) = \frac{5 \cdot 9 + 100}{10} = \frac{29}{2} = 14.5 \text{ and } s = 15.$$

Furthermore,

$$\#q(0.9) = \frac{5 + 9 \cdot 100}{10} = \frac{181}{2} = 90.5 \text{ and } t = 90.$$

We therefore have

$$\bar{T}(0.1) = \frac{\sum_{i=15}^{90} x_{(i)}}{90 - 15 + 1} = \frac{3 \cdot 11 + \cdots + 12 \cdot 1}{76} = \frac{453}{76} \approx 5.9605.$$

Example 4:
Determine the 10%-trimmed mean for the random gamma data.
Answer:
The tally table for the data is

4: 1	9: 2	13: 3	17: 2	24: 2
5: 4	10: 1	14: 2	18: 1	27: 1
8: 1	12: 1	15: 1	22: 2	31: 1

There are 25 observations in the sample.
By the traditional method,
$$\#q(0.1) = 0.1 \cdot 25 = 2.5 \text{ and } s = 3.$$
Furthermore,
$$\#q(0.9) = 0.9 \cdot 25 = 22.5 \text{ and } t = 22.$$
From this we get
$$\bar{T}(0.1) = \frac{\sum_{i=3}^{22} x_{(i)}}{22 - 3 + 1} = \frac{5 \cdot 3 + \cdots + 24 \cdot 1}{20} = \frac{53}{4} = 13.25.$$
By the restricted proper method, $\#(x_{\min}) = 1$ and $\#(x_{\max}) = 25$, as both are unique. Therefore,
$$\#q(0.1) = \frac{9 \cdot 1 + 25}{10} = 3.4 \text{ and } \#q(0.9) = \frac{1 + 9 \cdot 25}{10} = 22.6.$$
Wherefore, $s = 4$ and $t = 22$.
From the above we derive
$$\bar{T}(0.1) = \frac{\sum_{i=4}^{22} x_{(i)}}{22 - 4 + 1} = \frac{5 \cdot 2 + \cdots + 24 \cdot 1}{19} = \frac{260}{19} \approx 13.6842.$$

We note that for $p = 0.5$ we get the median for the 50%-trimmed mean.

Interquartile Mean

An alternative to Tukey's trimean is the 25%-trimmed mean. This is called the **interquartile mean** and is often denoted $\overline{T}_Q$. As this is determined by the quartiles, one additional method becomes available — the median-of-medians method.

Example 5:
Determine the interquartile mean for the two-success data.

Answer:
By the median-of-medians method,

$$\#(M) = \frac{1 + 100}{2} = 50.5 \text{ and } M = \frac{x_{(50)} + x_{(51)}}{2} = \frac{5+5}{2} = 5.$$

The highest index j such that $x_{(j)} \leq 5$ is $j = 57$; thus,

$$\#(Q_1) = \frac{1+57}{2} = 29 \text{ and } s = 30.$$

The smallest index j such that $x_{(j)} \geq 5$ is $j = 41$; whereby,

$$\#(Q_3) = \frac{41+100}{2} = 70.5 \text{ and } t = 70.$$

We conclude from the above that

$$\overline{T}_Q = \frac{\sum_{i=30}^{70} x_{(i)}}{70 - 30 + 1} = \frac{4 \cdot 11 + \cdots + 7 \cdot 4}{41} = \frac{211}{41} \approx 5.146.$$

By the traditional sample method,

$$\#(Q_1) = 0.25 \cdot 100 = 25 \text{ and } s = 26.$$

In addition,

$$\#(Q_3) = 0.75 \cdot 100 = 75 \text{ and } t = 75.$$

From this we obtain

$$\bar{T}_Q = \frac{\sum_{i=26}^{75} x_{(i)}}{75 - 26 + 1} = \frac{4 \cdot 15 + \cdots + 9 \cdot 1}{50} = \frac{134}{25} = 5.36.$$

In the restricted proper method, the effective addresses of the sample extremes are

$$\#(x_{\min}) = \frac{1+9}{2} = 5 \text{ and } \#(x_{\max}) = 100.$$

From this

$$\#(Q_1) = \frac{3 \cdot 5 + 100}{4} = 28.75 \text{ and } \#(Q_3) = \frac{5 + 3 \cdot 100}{4} = 76.25.$$

We conclude $s = 29$ and $t = 76$.
Using the above information,

$$\bar{T}_Q = \frac{\sum_{i=29}^{76} x_{(i)}}{76 - 29 + 1} = \frac{4 \cdot 12 + \cdots + 9 \cdot 2}{48} = \frac{265}{48} \approx 5.5208.$$

Example 6:
Determine the interquartile mean for the random gamma data.

Answer:
By the median-of-medians method,

$$\#(M) = \frac{1 + 25}{2} = 13 \text{ and } M = x_{(13)} = 13.$$

The highest index j such that $x_{(j)} \leq 13$ is $j = 13$; whereby,

$$\#(Q_1) = \frac{1 + 13}{2} = 7 \text{ and } s = 8.$$

5.9. TRIMMED MEANS

The lowest index j such that $x_{(j)} \geq 13$ is $j = 11$; thereby,

$$\#(Q_3) = \frac{11 + 25}{2} = 18 \text{ and } t = 18.$$

From the above we get

$$\bar{T}_Q = \frac{\sum_{i=8}^{18} x_{(i)}}{18 - 8 + 1} = \frac{9 \cdot 1 + \cdots + 17 \cdot 2}{11} = \frac{147}{11} \approx 13.3636.$$

By the traditional sample method,

$$\#(Q_1) = 0.25 \cdot 25 = 6.25 \text{ and } s = 7.$$

In addition,

$$\#(Q_3) = 0.75 \cdot 25 = 18.75 \text{ and } t = 18.$$

With the above information,

$$\bar{T}_Q = \frac{\sum_{i=7}^{18} x_{(i)}}{18 - 7 + 1} = \frac{9 \cdot 2 + \cdots 17 \cdot 2}{12} = 13.$$

The effective address for $x_{\min}$ is 1 and the effective address of $x_{\max}$ is 25 because each is unique. By the restricted sample method

$$\#(Q_1) = \frac{3 \cdot 1 + 25}{4} = 7 \text{ and } \#(Q_3) = \frac{1 + 3 \cdot 25}{4} = 19.$$

We conclude $s = 8$ and $t = 19$.

From the information above

$$\bar{T} = \frac{\sum_{i=8}^{19} x_{(i)}}{19 - 8 + 1} = \frac{9 \cdot 1 + \cdots + 18 \cdot 1}{12} = \frac{55}{4} = 13.75.$$

Exercises 5.9

1. One day while walking through a field you notice there seem to be a lot of stones lying on the ground. Curious about what the average size of a stone is in the field you randomly collect 30 stones. You weigh each stone and record its weight in grams. The following table shows the raw data.

110	126	75	110	126	94
90	112	111	139	131	92
102	103	96	75	67	101
104	111	100	65	92	93
106	104	93	115	109	95

   Determine $\bar{T}_3$, $\bar{T}(0.15)$ and $\bar{T}_Q$ for this data.

2. The scores, in the order graded, on a recent midterm examination in Statistics, a class with 30 students, are given below.

70	80	99	98	85	89
87	79	83	38	69	70
60	69	78	40	75	56
70	51	99	69	95	86
57	53	47	50	55	81

   Determine $\bar{T}_1$, $\bar{T}(0.1)$ and $\bar{T}_Q$ for this data.

3. Below is a sample of the monthly salary for 30 randomly selected people in mid-management in thousands of dollars.

3.08	3.18	3.13	7.41	3.73	3.05
4.17	3.58	3.36	3.27	3.36	4.74
3.32	3.61	4.26	3.02	3.45	4.06
6.73	4.72	3.13	3.15	3.70	3.59
3.12	3.61	5.03	3.53	3.20	3.32

   Determine $\bar{T}_2$, $\bar{T}(0.2)$ and $\bar{T}_Q$ for this data.

5.10. WINSORIZED MEANS

4. A pair of dice, one red and one green, is rolled in the following manner. On the first toss both dice are tossed. Then the red die is tossed while the green die keeps its value to get the second sum. Then the green die is tossed while the red die keeps its value. You continue tossing alternately the red then the green, die and generate the following table.

8	11	8	6	10	9	9
7	2	5	10	11	8	5
3	6	9	7	7	9	10
7	3	3	7	9	8	7
5	4	8	11	11	10	10

Determine $\bar{T}_1$, $\bar{T}(0.1)$ and $\bar{T}_Q$ for this data.

5. The following random sample of 72 observations comes from a gamma distribution, so $\mu_{min} = 0$ and μ_{max} has no upper limit. The raw data are

2.15	1.92	1.92	2.20	3.67	4.55	3.32	2.71	4.87
2.09	1.13	1.09	2.12	4.89	3.03	3.17	4.74	4.44
4.61	2.78	2.97	2.78	2.88	3.66	2.55	4.40	3.54
2.18	3.43	3.84	4.60	1.76	2.85	2.27	1.38	1.24
4.22	1.67	4.51	3.50	6.69	5.92	6.42	0.95	4.57
3.10	3.26	2.95	2.45	1.95	2.06	3.10	2.03	4.98
0.86	2.61	1.92	2.78	1.01	2.64	3.14	0.86	3.68
2.71	3.06	4.33	4.63	2.47	2.92	1.38	2.81	5.71

Determine T_3, $T(0.20)$ and T_Q for this data.

5.10 Winsorized Means

CONSIDER THE sorted observations in pairs and take the Walsh sums of consecutive pairs as you head toward the middle observations. If the data are balanced then there will not be much variation in the values of these sums. Even in the presence of outliers, only the first sums may differ drastically.

Motivation 1:
Below we show the sorted data for the two-successes experiment.

2	2	2	2	2	2	2	2	2	3
3	3	3	3	3	3	3	3	3	3
3	3	3	3	3	4	4	4	4	4
4	4	4	4	4	4	4	4	4	4
5	5	5	5	5	5	5	5	5	5
5	5	5	5	5	5	5	6	6	6
6	6	6	6	6	6	7	7	7	7
8	8	8	8	9	9	9	9	9	10
10	10	10	10	10	11	11	11	11	12
12	14	14	14	14	14	15	16	17	28

The 50 Walsh sums of pairs working from the ends to the middle are

15.0	9.5	9.0	8.5	8.0	8.0	8.0	8.0	8.0	7.5
7.5	7.0	7.0	7.0	7.0	6.5	6.5	6.5	6.5	6.5
6.5	6.0	6.0	6.0	6.0	6.5	6.0	6.0	6.0	6.0
5.5	5.5	5.5	5.5	5.0	5.0	5.0	5.0	5.0	5.0
5.5	5.5	5.5	5.0	5.0	5.0	5.0	5.0	5.0	5.0

After the first few they do not vary much in value.

r-Winsorized Means

THE TWO-SUCCESSES DATA is not particularly well-balanced, either and yet the motivational example still shows that the Walsh sums settle down quickly. This means that instead of eliminating the outliers, we can replace pairs of eliminated data with pairs of the last

5.10. WINSORIZED MEANS

kept values and we will get a smoother estimate for the mean rather than just ignoring the contribution of the ends altogether.

The r-**Winsorized mean**, denoted $\overline{W}_r$, replaces the first r sorted observations with $x_{(r+1)}$ and the last r sorted observations with $x_{(n-r)}$ and then calculates the mean.

$$\overline{W}_r = \frac{r \cdot (x_{(r+1)} + x_{(n-r)}) + \sum_{i=r+1}^{n-r} x_{(i)}}{n}.$$

If we have already calculated the r-trimmed mean then

$$\overline{W}_r = \frac{(n - 2 \cdot r) \cdot \overline{T}_r}{n} + \frac{r \cdot (x_{(r+1)} + x_{(n-r)})}{n}.$$

Example 1:
Calculate $\overline{W}_1$, $\overline{W}_2$ and $\overline{W}_9$ for the two-successes data directly. Then use the results of Example 1, §5.9, which started on page 243 to verify you get the same result from the corresponding r-trimmed mean.

Answer:
The 1-Winsorized mean is

$$\overline{W}_1 = \frac{2 + 17 + \sum_{i=1}^{99} x_{(i)}}{100} = \frac{2 \cdot 9 + \cdots + 17 \cdot 2}{100} = \frac{633}{100} = 6.33.$$

The 1-trimmed mean is $\overline{T}_1 = 307/49$. By the modifying formula,

$$\overline{W}_1 = \frac{(100 - 2) \cdot (307/49)}{100} + \frac{2 + 19}{100} = \frac{614 + 19}{100} = \frac{633}{100} = 6.33 \checkmark$$

The 2-Winsorized mean is

$$\bar{W}_2 = \frac{2 \cdot (2 + 16) + \sum_{i=3}^{98} x_{(i)}}{100} = \frac{2 \cdot 9 + \cdots + 16 \cdot 3}{100} = \frac{631}{100} = 6.31.$$

The 2-trimmed mean is $\bar{T}_2 = 595/96$. By the modifying formula,

$$\bar{W}_2 = \frac{(100 - 2 \cdot 2) \cdot (595/96)}{100} + \frac{2 \cdot (2 + 16)}{100}$$

$$= \frac{595 + 36}{100} = \frac{631}{100} = 6.31 \checkmark$$

The 9-Winsorized mean is

$$\bar{W}_9 = \frac{9 \cdot (3 + 12) + \sum_{i=10}^{91} x_{(i)}}{100} = \frac{3 \cdot 25 + \cdots + 12 \cdot 11}{100} = \frac{123}{20} = 6.15.$$

The 9-trimmed mean is $\bar{T}_9 = 240/41$. By the modifying formula,

$$\bar{W}_9 = \frac{(100 - 2 \cdot 9) \cdot (240/41)}{100} + \frac{9 \cdot (3 + 12)}{100}$$

$$= \frac{480 + 135}{100} = \frac{123}{20} = 6.15 \checkmark$$

Example 2:
Determine the 1-Winsorized mean for the random gamma data. Verify you get the same result from the result of Example 2, §5.9, which started on page 245, and the modifying formula.

Answer:
The 1-Winsorized mean is

$$\bar{W}_1 = \frac{5 + 27 + \sum_{i=2}^{24} x_{(i)}}{25} = \frac{5 \cdot 5 + \cdots + 27 \cdot 2}{25} = \frac{353}{25} = 14.12.$$

5.10. WINSORIZED MEANS

The 1-trimmed mean is $\overline{T}_1 = 321/23$. By the modifying formula,

$$\overline{W}_1 = \frac{(25-2) \cdot (321/23)}{25} + \frac{5+27}{25} = \frac{321+32}{25} = \frac{353}{25} = 14.12 \checkmark$$

p-Winsorized Means

THERE ARE TWO ways to perform trimming; hence, two ways to perform Winsorizing. The theory is basically the same — replace the trimmed sections with Walsh sums to bring the count back up. However, the p-trimmed mean does not always trim off the same number of elements from each end. Because of this, the modification formula for creating a p-Winsorized mean from its corresponding p-trimmed mean does not always exactly match the result of calculating the p-Winsorized mean directly — although they are always close to each other. The **p-Winsorized mean**, denoted $\overline{W}(p)$, replaces all elements below $q(p)$ with $q(p)$ and all elements above $q(1-p)$ with $q(1-p)$ and leaves all other elements unchanged in calculating the mean. Therefore,

$$\overline{W}(p) = \frac{\sum_{i=1}^{\lfloor \#q(p) \rfloor} q(p) + \sum_{i=\lfloor \#q(p) \rfloor+1}^{\lfloor \#q(1-p) \rfloor} x_{(i)} + \sum_{i=\lfloor \#q(1-p) \rfloor+1}^{n} q(1-p)}{n}$$

If the p-trimmed mean has been previously calculated then there is an approximation formula for the p-Winsorized mean that can be created from it.

$$\overline{W}(p) \approx (1 - 2 \cdot p) \cdot \overline{T}(p) + p \cdot \big(q(p) + q(1-p)\big).$$

Example 3:
Determine the 10%-Winsorized mean for the two successes data. Then use the approximation formula and the estimate of $\overline{T}(0.1)$ from Example 3, §5.9, which started on page 246 to again find $\overline{W}(0.1)$.

Answer:
   In the traditional sample method,
$$\#q(0.1) = 0.1 \cdot 100 = 10 \text{ and } q(0.1) = \frac{x_{(10)} + x_{(11)}}{2} = \frac{3+3}{2} = 3.$$

Furthermore,
$$\#q(0.9) = 0.9 \cdot 100 = 90 \text{ and } q(0.9) = \frac{x_{(90)} + x_{(91)}}{2} = \frac{12+12}{2} = 12.$$

This makes
$$\overline{W}(0.1) = \frac{\sum_{i=1}^{10} 3 + \sum_{i=11}^{90} x_{(i)} + \sum_{i=91}^{100} 12}{100}$$
$$= \frac{3 \cdot 25 + \cdots + 12 \cdot 11}{100} = \frac{123}{20} = 6.15.$$

The 10%-trimmed mean is $\overline{T}(0.1) = 93/16$. The approximation formula gives
$$\overline{W}(0.1) \approx \frac{4}{5} \cdot \frac{93}{16} + \frac{1}{10} \cdot (3+12) = \frac{93}{20} + \frac{3}{2} = \frac{123}{20} = 6.15.$$

The same number of elements were trimmed from each end, so the approximation formula is exact.
   The effective position of $x_{\min} = 2$ is $\#(2) = (1+9)/2 = 5$. The effective position of $x_{\max} = 28$ is $\#(28) = 100$. Therefore, the restricted proper method gives
$$\#q(0.1) = \frac{9 \cdot 5 + 100}{10} = 14.5 \text{ and } \#q(0.9) = \frac{5 + 9 \cdot 100}{10} = 90.5$$

5.10. WINSORIZED MEANS

For $q(0.1)$, $x_{(14)} = 3$ and $\#(3) = (10+25)/2 = 17.5 > 14.5$. Wherefore, $y_2 = 3$, $m_2 = 17.5$, $y_1 = 2$ and $m_1 = 5$. By interpolation,

$$q(0.1) = \frac{2 \cdot (17.5 - 14.5) + 3 \cdot (14.5 - 5)}{17.5 - 5} = \frac{6 + 28.5}{12.5} = \frac{69}{25} = 2.76.$$

For $q(0.9)$, $x_{(90)} = 12$ and $\#(12) = (90+91)/2 = 90.5 = \#q(0.9)$, so $q(0.9) = 12$. This means that

$$\overline{W}(0.1) = \frac{\sum_{i=1}^{14} \frac{69}{25} + \sum_{i=15}^{90} x_{(i)} + \sum_{i=91}^{100} 12}{100}$$

$$= \frac{(966/25) + 453 + 120}{100} = \frac{15291}{2500} = 6.1164.$$

The 10%-trimmed mean is $T(0.1) = 453/76$. From the approximation formula we have

$$\overline{W}(0.1) \approx \frac{4}{5} \cdot \frac{453}{76} + \frac{1}{10} \cdot \left(\frac{69}{25} + 12 \right) = \frac{453}{95} + \frac{369}{250} = \frac{29661}{4750} \approx 6.2444.$$

More elements were trimmed from the bottom than the top, so the approximation formula will give too large a result.

Example 4:
Determine the 10%-Winsorized mean for the random gamma data. Then use the approximation formula and the estimate of $T(0.1)$ from Example 4, §5.9, which started on page 248 to again find $\overline{W}(0.1)$.

Answer:
In the traditional sample method,

$$\#q(0.1) = 0.1 \cdot 25 = 2.5; q(0.1) = x_{(3)} = 5 \text{ and } s = 3.$$

In addition,

$$\#q(0.9) = 0.9 \cdot 25 = 22.5; q(0.9) = x_{(23)} = 24 \text{ and } t = 22.$$

We therefore have

$$\bar{W}(0.1) = \frac{\sum_{i=1}^{2} 5 + \sum_{i=3}^{22} x_{(i)} + \sum_{i=23}^{25} 24}{25} = \frac{5 \cdot 5 \cdots + 24 \cdot 4}{25} = \frac{347}{25} = 13.88.$$

The 10%-trimmed mean is $\bar{T}(0.1) = 53/4$. By the approximation formula we get

$$\bar{W}(0.1) \approx \frac{4}{5} \cdot \frac{53}{4} + \frac{1}{10} \cdot (5 + 24) = \frac{53}{5} + \frac{29}{10} = \frac{27}{2} = 13.5.$$

More elements were trimmed from the top than the bottom, so the approximation formula will give too small a result.

Both $x_{\min} = 4$ and $x_{\max} = 31$ are unique; thus,

$$\#q(0.1) = \frac{9 \cdot 1 + 25}{10} = 3.4 \text{ and } s = 4.$$

Also,

$$\#q(0.9) = \frac{1 + 9 \cdot 25}{10} = 22.6 \text{ and } t = 22.$$

For $q(0.1)$, $x_{(3)} = 5$ and $\#(5) = (2+5)/2 = 3.5 > 3.4$. We conclude $y_2 = 5$, $m_2 = 3.5$, $y_1 = 4$ and $m_1 = 1$. By interpolation,

$$q(0.1) = \frac{4 \cdot (3.5 - 3.4) + 5 \cdot (3.4 - 1)}{3.5 - 1} = \frac{0.4 + 12}{2.5} = \frac{124}{25} = 4.96.$$

For $q(0.9)$, $x_{(22)} = 24$ and $\#(24) = (22+23)/2 = 22.5 < 22.6$. We conclude $y_1 = 24$, $m_1 = 22.5$, $y_2 = 27$ and $m_2 = 24$. By interpolation,

$$q(0.9) = \frac{24 \cdot (24 - 22.6) + 27 \cdot (22.6 - 22.5)}{24 - 22.5}$$
$$= \frac{33.6 + 2.7}{1.5} = \frac{121}{5} = 24.2.$$

5.10. WINSORIZED MEANS

We thus have

$$\overline{W}(0.1) = \frac{\sum_{i=1}^{3} \frac{124}{25} + \sum_{i=4}^{22} x_{(i)} + \sum_{i=23}^{25} \frac{121}{5}}{25}$$

$$= \frac{(372/25) + 5 \cdot 1 + \cdots + 24 \cdot 1 + (363/5)}{25}$$

$$= \frac{8562}{625} = 13.6992.$$

The 10%-trimmed mean is $\overline{T}(0.1) = 260/19$. The approximation formula gives

$$\overline{W}(0.1) \approx \frac{4}{5} \cdot \frac{260}{19} + \frac{1}{10} \cdot \left(\frac{124}{25} + \frac{121}{5} \right) = \frac{208}{19} + \frac{729}{250} = \frac{65851}{4750} \approx 13.8634.$$

Even though three elements were trimmed from each end, because less than 20 elements were involved in T, the result is not exact.

The only time that the approximation formula will give the same result as the direct formula is when exactly $100(1 - 2p)\%$ of the observations are involved in the calculation of $\overline{T}(p)$ and the same number of elements were trimmed from each end to get it.

Winsorized Interquartile Mean

THE MOST COMMON p-value on a p-Winsorized mean is $p = 25\%$. This is the **Winsorized interquartile mean** and is denoted $\overline{W}_Q$. It corresponds to $\overline{T}_Q$ in the same way that $\overline{W}(p)$ corresponds to $\overline{T}(p)$ for any other value of p. Of course, being based upon the quartiles it is possible to use the median-of-medians method as well as the traditional sample and restricted proper methods.

$$\overline{W}_Q = \frac{\sum_{i=1}^{\lfloor \#(Q_1) \rfloor} Q_1 + \sum_{i=\lfloor \#(Q_1) \rfloor+1}^{\lfloor \#(Q_3) \rfloor} x_{(i)} + \sum_{i=\lfloor \#(Q_3) \rfloor+1}^{n} Q_3}{n}.$$

Furthermore, it can be approximated as

$$\overline{W}_Q \approx \frac{\overline{T}_Q}{2} + \frac{Q_1 + Q_3}{4}.$$

Example 5:
Determine the Winsorized interquartile mean for the two-successes data.

Answer:
With the median-of-medians method, $\#(M) = (1 + 100)/2 = 50.5$, so $M = 5$ as $x_{(50)} = x_{(51)} = 5$. The highest j such that $x_{(j)} \leq M$ is $j = 57$ as $x_{(57)} = 5$ and $x_{(58)} = 6$. Therefore,

$$\#(Q_1) = \frac{1 + 57}{2} = 29 \text{ and } Q_1 = x_{(29)} = 4.$$

The lowest j such that $x_{(j)} \geq M$ is $j = 41$ as $x_{(41)} = 5$ and $x_{(40)} = 4$. Wherefore,

$$\#(Q_3) = \frac{41 + 100}{2} = 70.5 \text{ and } Q_3 = \frac{x(70) + x_{(71)}}{2} = \frac{7 + 8}{2} = 7.5.$$

Herefrom,

$$\overline{W}_Q = \frac{\sum_{i=1}^{29} 4 + \sum_{i=30}^{70} x_{(i)} + \sum_{i=71}^{100} 7.5}{100}$$
$$= \frac{4 \cdot 40 + \cdots + 7 \cdot 4 + 7.5 \cdot 30}{100} = \frac{138}{25} = 5.52.$$

5.10. WINSORIZED MEANS

The interquartile mean by the median-of-medians method is $\overline{T}_Q = 211/41$. The approximation formula gives

$$\overline{W}_Q \approx \frac{211/41}{2} + \frac{4+7.5}{4} = \frac{211}{82} + \frac{23}{8} = \frac{1787}{328} \approx 5.4482.$$

Less than half the data were contained in the formula for $\overline{T}_Q$, so the quartiles affect the result more than they should. More were trimmed from the top, so the result is too small.

With the traditional sample method

$$\#(Q_1) = \frac{100}{4} = 25 \text{ and } Q_1 = \frac{x_{(25)} + x_{(26)}}{2} = \frac{3+4}{2} = 3.5.$$

Furthermore,

$$\#(Q_3) = 3\#(Q_1) = 75 \text{ and } Q_3 = \frac{x_{(75)} + x_{(76)}}{2} = \frac{9+9}{2} = 9.$$

Therefrom,

$$\overline{W}_Q = \frac{\sum_{i=1}^{25} 3.5 + \sum_{i=26}^{75} x_{(i)} + \sum_{i=76}^{100} 9}{100}$$
$$= \frac{3.5 \cdot 25 + 4 \cdot 15 + \cdots + 9 \cdot 26}{100}$$
$$= \frac{1161}{200} = 5.805.$$

The interquartile mean from the traditional sample method is $\overline{T}_Q = 134/25$. That, with the approximation formula gives

$$\overline{W}_Q \approx \frac{134/25}{2} + \frac{3.5+9}{4} = \frac{67}{25} + \frac{25}{8} = \frac{1161}{200} = 5.805.$$

Exactly half the data were used for $\overline{T}_Q$ and the same number of elements were trimmed from each end. Thus, we get equality for the approximation formula.

By the restricted proper method

$$\#(x_{\min}) = \frac{1+9}{2} = 5 \text{ and } \#(x_{\max}) = 100.$$

Whence,

$$\#(Q_1) = \frac{3 \cdot 5 + 100}{4} = 28.75 \text{ and } \#(Q_3) = \frac{5 + 3 \cdot 100}{4} = 76.25.$$

To obtain Q_1 we note that $x_{(28)} = 4$ and $\#(4) = (26+40)/2 = 33 > 28.75$. Thence, $y_2 = 4$, $m_2 = 33$, $y_1 = 3$ and $m_1 = (10+25)/2 = 17.5$. By interpolation,

$$Q_1 = \frac{3 \cdot (33 - 28.75) + 4 \cdot (28.75 - 17.5)}{33 - 17.5} = \frac{12.75 + 45}{15.5} = \frac{231}{62}.$$

To get Q_3 we observe $x_{(76)} = 9$ and $\#(9) = (75+79)/2 = 77 > 76.25$. Hence, $y_2 = 9$, $m_2 = 77$, $y_1 = 8$ and $m_1 = (71+74)/2 = 72.5$. Using linear interpolation,

$$Q_3 = \frac{8 \cdot (77 - 76.25) + 9 \cdot (76.25 - 72.5)}{77 - 72.5} = \frac{6 + 33.75}{4.5} = \frac{53}{6}.$$

Wherefrom,

$$\overline{W}_Q = \frac{\sum_{i=1}^{28} \frac{231}{62} + \sum_{i=29}^{76} x_{(i)} + \sum_{i=77}^{100} \frac{53}{6}}{100}$$
$$= \frac{(3234/31) + 265 + 212}{100}$$
$$= \frac{18021}{3100} \approx 5.8132.$$

The interquartile mean obtained from the restricted proper method is $\overline{T}_Q = 265/48$. Placing this into the approximation formula

$$\overline{W}_Q \approx \frac{265/48}{2} + \frac{(231/62) + (53/6)}{4} = \frac{265}{96} + \frac{292}{93} = \frac{5853}{992} \approx 5.9002.$$

5.10. WINSORIZED MEANS

Less than half the data lies between the quartiles, so the effect of the quartiles was magnified. More were trimmed from the bottom, so the result is too large.

Example 6:
Determine the Winsorized interquartile mean for the random gamma data. Then use the approximation formula to estimate W_Q from the interquartile mean calculated in Example 6, §5.9, which started on page 250.

Answer:
For the median-of-medians method, $\#(M) = (1+25)/2 = 13$ and $M = x_{(13)} = 13$. The highest j such that $x_{(j)} \leq M$ is $j = 13$ as $x_{(13)} = 13$ and $x_{(14)} = 14$. Therefore,

$$\#(Q_1) = \frac{1+13}{2} = 7 \text{ and } Q_1 = x_{(7)} = 9.$$

The lowest j such that $x_{(j)} \geq M$ is $j = 11$ as $x_{(11)} = 13$ and $x_{(10)} = 12$. Wherefrom,

$$\#(Q_3) = \frac{11+25}{2} = 18 \text{ and } Q_3 = x_{(18)} = 17.$$

Thereby,

$$\overline{W}_Q = \frac{\sum_{i=1}^{7} 9 + \sum_{i=8}^{18} x_{(i)} + \sum_{i=19}^{25} 17}{25} = \frac{9 \cdot 8 + \cdots + 17 \cdot 9}{25} = \frac{329}{25} = 13.16.$$

The interquartile mean obtained with the median-of-medians method is $T_Q = 147/11$. The approximation formula gives

$$\overline{W}_Q \approx \frac{147/11}{2} + \frac{9+17}{4} = \frac{147}{22} + \frac{13}{2} = \frac{145}{11} \approx 13.1818.$$

Because the interquartile mean involved less than half of the data the quartiles had an exaggerated effect, even though the same number of elements were trimmed from each end.

By the traditional sample method,

$$\#(Q_1) = \frac{25}{4} = 6.25 \text{ and } Q_1 = x_{(7)} = 9.$$

In addition,

$$\#(Q_3) = 3\#(Q_1) = 18.75 \text{ and } Q_3 = x_{(19)} = 18.$$

Whereby,

$$\overline{W}_Q = \frac{\sum_{i=1}^{6} 9 + \sum_{i=7}^{18} x_{(i)} + \sum_{i=19}^{25} 18}{25}$$
$$= \frac{9 \cdot 8 + \cdots + 17 \cdot 2 + 18 \cdot 7}{25}$$
$$= \frac{336}{25} = 13.44.$$

The interquartile mean using the traditional sample method is $\overline{T}_Q = 13$. Placing this into the approximation formula gives

$$\overline{W}_Q \approx \frac{13}{2} + \frac{9+18}{4} = \frac{13}{2} + \frac{27}{4} = \frac{53}{4} = 13.25.$$

More items were trimmed from the top than the bottom, so the estimate is too small. Also, less than half of the data was included within the calculation of $\overline{T}_Q$ so the effect of the quartiles is slightly larger than it should be.

Both $x_{\min} = 4$ and $x_{\max} = 31$ are unique. The restricted therefore calculates

$$\#(Q_1) = \frac{3 \cdot 1 + 25}{4} = 7 \text{ and } \#(Q_3) = \frac{1 + 3 \cdot 25}{4} = 19.$$

5.10. WINSORIZED MEANS

Considering Q_1 we see $x_{(7)} = 9$ and $\#(9) = (7+8)/2 = 7.5 > 7$. Thus, $y_2 = 9$, $m_2 = 7.5$, $y_1 = 8$ and $m_1 = 6$. From linear interpolation we find

$$Q_1 = \frac{8 \cdot (7.5 - 7) + 9 \cdot (7 - 6)}{7.5 - 6} = \frac{4 + 9}{1.5} = \frac{26}{3}.$$

For Q_3 we note that $x_{(19)} = 18$ is unique, so $Q_3 = 18$ as $\#(18) = \#(Q_3)$. Hereby,

$$W_Q = \frac{\sum_{i=1}^{7} \frac{26}{3} + \sum_{i=8}^{19} x_{(i)} + \sum_{i=20}^{25} 18}{25}$$
$$= \frac{(182/3) + 165 + 108}{25}$$
$$= \frac{1001}{75} \approx 13.3467.$$

The restricted proper method calculated the interquartile mean to be $T_Q = 55/4$; thus, the approximation formula yields

$$W_Q \approx \frac{55/4}{2} + \frac{(26/3) + 18}{4} = \frac{55}{8} + \frac{20}{3} = \frac{325}{24} \approx 13.5417.$$

More elements were trimmed from the bottom than the top, so the estimate is larger than the actual value.

Exercises 5.10

1. One day while walking through a field you notice there seem to be a lot of stones lying on the ground. Curious about what the average size of a stone is in the field you randomly collect 30 stones. You

weigh each stone and record its weight in grams. The following table shows the raw data.

110	126	75	110	126	94
90	112	111	139	131	92
102	103	96	75	67	101
104	111	100	65	92	93
106	104	93	115	109	95

Determine $\bar{W}_3$, $\bar{W}(0.15)$ and $\bar{W}_Q$ for this data. Also determine the approximations based upon the results of the corresponding exercise from §5.9.

2. The scores, in the order graded, on a recent midterm examination in Statistics, a class with 30 students, are given below.

70	80	99	98	85	89
87	79	83	38	69	70
60	69	78	40	75	56
70	51	99	69	95	86
57	53	47	50	55	81

Determine $\bar{W}_1$, $\bar{W}(0.1)$ and $\bar{W}_Q$ for this data. Also determine the approximations based upon the results of the corresponding exercise from §5.9.

3. Below is a sample of the monthly salary for 30 randomly selected people in mid-management in thousands of dollars.

3.08	3.18	3.13	7.41	3.73	3.05
4.17	3.58	3.36	3.27	3.36	4.74
3.32	3.61	4.26	3.02	3.45	4.06
6.73	4.72	3.13	3.15	3.70	3.59
3.12	3.61	5.03	3.53	3.20	3.32

Determine $\bar{W}_2$, $\bar{W}(0.2)$ and $\bar{W}_Q$ for this data. Also determine the approximations based upon the results of the corresponding exercise from §5.9.

5.10. WINSORIZED MEANS

4. A pair of dice, one red and one green, is rolled in the following manner. On the first toss both dice are tossed. Then the red die is tossed while the green die keeps its value to get the second sum. Then the green die is tossed while the red die keeps its value. You continue tossing alternately the red then the green, die and generate the following table.

8	11	8	6	10	9	9
7	2	5	10	11	8	5
3	6	9	7	7	9	10
7	3	3	7	9	8	7
5	4	8	11	11	10	10

Determine $\overline{W}_1$, $\overline{W}(0.1)$ and $\overline{W}_Q$ for this data. Also determine the approximations based upon the results of the corresponding exercise from §5.9.

5. The following random sample of 72 observations comes from a gamma distribution, so $\mu_{\min} = 0$ and $\mu_{\max}$ has no upper limit. The raw data are

2.15	1.92	1.92	2.20	3.67	4.55	3.32	2.71	4.87
2.09	1.13	1.09	2.12	4.89	3.03	3.17	4.74	4.44
4.61	2.78	2.97	2.78	2.88	3.66	2.55	4.40	3.54
2.18	3.43	3.84	4.60	1.76	2.85	2.27	1.38	1.24
4.22	1.67	4.51	3.50	6.69	5.92	6.42	0.95	4.57
3.10	3.26	2.95	2.45	1.95	2.06	3.10	2.03	4.98
0.86	2.61	1.92	2.78	1.01	2.64	3.14	0.86	3.68
2.71	3.06	4.33	4.63	2.47	2.92	1.38	2.81	5.71

Determine $\overline{W}_3$, $\overline{W}(0.20)$ and $\overline{W}_Q$ for this data. Also determine the approximations based upon the results of the corresponding exercise from §5.9.

Chapter 6

Data Variability

Dispersion measures the spread of the data, the spacing between observations, or the deviation of the data from a fixed point. In this chapter we look at several statistics to estimate variability.

6.1 Range & Interquartile Range

WE BEGIN WITH spread and revisit two statistics we have seen before, the range and the interquartile range. We will then look at a few statistics based upon these estimators.

The Range

THE SMALLEST possible value in the population is $\mu_{\min}$. In the sample this parameter is estimated by the statistic $x_{\min}$. The basic difference between them is that $\mu_{\min}$ may be unknown or even nonexistent, but x_{min} is always there. It may vary from sample to sample, but there always is one.

The largest possible value in the population is μ_{max}. In the sample this parameter is estimated by the statistic x_{max}. Just as with μ_{min}, μ_{max} may be unknown or even nonexistent, but x_{max} is always there. Again, it may vary from sample to sample, but there always is one.

When both μ_{min} and μ_{max} exist and are known values then the spread between these values is termed the **population range** and is denoted $\mathcal{R}$. Thus,
$$\mathcal{R} = \mu_{max} - \mu_{min}.$$
The statistic corresponding to this parameter is the **sample range** and is denoted R. Hence,
$$R = x_{max} - x_{min}.$$
The population parameter will not exist if even one of the parameters μ_{min} or μ_{max} is unknown or nonexistent. The sample statistic, however, always exists, but may vary from sample to sample.

A statistic is **robust** if the value of the statistic is not influenced much by the presence of outliers. Clearly the range cannot be a robust measurement as its value is derived from the most extreme measurements. Still, when the population parameters are known — or at least definable — then the range does not seem to vary much between samples of the same size from the same population. Therefore, this can be a useful statistic to look at.

The Interquartile Range

THERE ARE MANY ways to calculate the quartiles. We begin with a quick review of the methods.

6.1. RANGE & INTERQUARTILE RANGE

The averaging, nearest and estimation methods can be used when the population minimum and maximum are not present in the data. The estimation method always lies exactly halfway between the averaging and nearest methods for the quartiles. The effective address of the quartiles are

$$\#(Q_1) = \frac{1+n}{4} \text{ and } \#(Q_3) = 3\#(Q_1).$$

If the effective position is an integer, then all methods take that sorted datum for the quartile; otherwise, the averaging method uses the arithmetic average of between the two observations whose sorted position is on either side of this value, the nearest method uses the one whose position is closer, and the estimation method is the arithmetic average of the other two methods — that is, it uses linear interpolation between the two closest sorted data.

On the rare instance that the effective addresses of the quartiles are exactly halfway between integers, then the even variation will use the observation in the nearest even-numbered sorted position. In this case, the estimation method method is not necessarily the arithmetic average between the averaging and nearest methods.

The regular proper method can be used in the above case, but can also be used when one or both of the population extremes is an observed value in the sample. The effective address for the quartiles are

$$\#(Q_1) = \frac{3\#(\mu_{\min}) + \#(\mu_{\max})}{4}$$

and

$$\#(Q_3) = \frac{\#(\mu_{\min}) + 3\#(\mu_{\max})}{4}.$$

If we let $\#Q$ be the effective address of the quartile sought, then $m = \lfloor \#Q \rfloor$ and we calculate $\#(x_{(m)})$. Using Q for the quartile sought we have three cases.

If $\#(x_{(m)}) = \#Q$ then $Q = x_{(m)}$.

If $\#(x_{(m)}) < \#Q$ then $y_1 = x_{(m)}$ and $m_1 = \#(x_{(m)})$. We set y_2 to be the next higher distinct observation and $m_2 = \#(y_2)$.

If $\#(x_{(m)}) > \#Q$ then $y_2 = x_{(m)}$ and $m_2 = \#(x_{(m)})$. We set y_1 to be the next lower distinct observation and $m_1 = \#(y_1)$.

If the value of Q has not yet been determined then one of the last two cases occurred. In either event, $y_1 < Q < y_2$ and $m_1 < \#Q < m_2$. We use linear interpolation to determine Q.

$$\begin{aligned} Q &= \frac{y_1 \cdot (m_2 - \#Q) + y_2 \cdot (\#Q - m_1)}{m_2 - m_1} \\ &= y_1 + \frac{(\#Q - m_1) \cdot (y_2 - y_1)}{m_2 - m_1}. \end{aligned}$$

Many people prefer a method which restricts the quartiles to the information in the sample, treating $x_{\min}$ as $\mu_{\min}$ and $x_{\max}$ as $\mu_{\max}$. For this there are three methods.

The median-of-medians method begins by determining the median. The effective address of the median is

$$\#(M) = \frac{1+n}{2}.$$

If $n = 2m-1$ then $M = x_{(m)}$. If $n = 2m$ then $M = (x_{(m)} + x_{(m+1)})/2$.

We now determine the largest integer k such that $x_{(k)} \leq M$ and if $k < n$ then $x_{(k+1)} > M$. We calculate the effective address of Q_1 as

$$\#(Q_1) = \frac{1+k}{2}.$$

If $m = \lfloor \#(Q_1) \rfloor = \#(Q_1)$ then $Q_1 = x_{(m)}$; otherwise, $Q_1 = (x_{(m)} + x_{(m+1)})/2$.

We next determine the smallest integer j such that $x_{(j)} \geq M$ and if $j > 1$ then $x_{(j-1)} < M$. The effective address of Q_3 is

$$\#(Q_3) = \frac{j+n}{2}.$$

If $m = \lfloor \#(Q_3) \rfloor = \#(Q_3)$ then $Q_3 = x_{(m)}$; otherwise, $Q_3 = (x_{(m)} + x_{(m+1)})/2$.

Each quartile is calculated as a median for some subset of the data; whence, the name of the method.

The traditional sample method calculates the effective address of the quartiles as

$$\#(Q_1) = \frac{n}{4} \text{ and } \#(Q_3) = 3\#(Q_1).$$

Let Q be the quartile sought and $\#Q$ its effective address. If $m = \lfloor \#Q \rfloor = \#Q$ then $Q = (x_{(m)} + x_{(m+1)})/2$; otherwise, $Q = x_{(m+1)}$.

If neither $\mu_{\min}$ nor $\mu_{\max}$ is an observed value in the sample then the quartiles obtained from the traditional sample method are exactly the same as given by the averaging method.

The restricted proper method is virtually identical to the regular proper method. The difference is that μ is replaced by x; thence,

$$\#(Q_1) = \frac{3\#(x_{\min}) + \#(x_{\max})}{4}$$

and

$$\#(Q_3) = \frac{\#(x_{\min}) + 3\#(x_{\max})}{4}.$$

We now proceed in the same manner. When $x_{\min}$ and $x_{\max}$ are unique, and either both are equal to the population parameters or neither is, then the results of the restricted proper method will be identical to the results from the regular proper method.

When the data are grouped, the effective address for the quartiles by the grouped method are

$$\#(Q_1) = \frac{n}{4} \text{ and } \#(Q_3) = 3\#(Q_1),$$

just like the traditional sample method. The difference is that we do not have the individual observations to determine Q, the quartile being sought. Again we let $\#Q$ be the effective address of the sought quartile. Suppose there are g group intervals, that interval i goes from t_{i-1} to t_i $(i = 1, \ldots, g)$ and that f_i is the number of observations falling in interval i. Then, we determine j such that

$$\text{offset} = \#Q - \sum_{i=1}^{j-1} f_i \leq f_j.$$

From this,

$$Q = t_{j-1} + \frac{\text{offset} \cdot (t_j - t_{j-1})}{f_j}.$$

Regardless of the manner in which the quartiles are calculated, the **interquartile range**, denoted $\boldsymbol{R_Q}$, is defined as

$$R_Q = Q_3 - Q_1.$$

This is the spread for the core of the data. The **core** is the middle fifty percent.

The quartiles will exist for the population even when the population extremes are unknown or nonexistent. However, one may not be able to calculate them directly even when the population extremes are known. When we get into random variables and probability distributions we will discuss some methods which may be used. Until then, we will note that the sample quartiles are usually good estimators and do not vary much between samples taken from the same population. The interquartile range is not affected by the presence or absence of outliers and is therefore considered a robust statistic.

The Clustering Factor

THE REST OF THIS section is devoted to statistics based upon the range and interquartile range. The first statistic we will look at is the clustering factor. The **clustering factor** is the ratio of the interquartile range to the range and is denoted by $\mathbb{C}$.

$$\mathbb{C} = \frac{R_Q}{R}$$

This statistic answers the question: *How concentrated is the core?* If the core is large in comparison to the entire sample then the data are considered **dispersed away from the median** — spread out. If the core is small with respect to the entire sample then the data are **clustered near the median**. In between values are neither clustered nor disperse. The cutoff points are

$$\text{If } \begin{cases} \mathbb{C} \geq 0.7 & \text{dispersed from median} \\ 0.3 < \mathbb{C} < 0.7 & \text{neither disperse nor clustered} \\ \mathbb{C} \leq 0.3 & \text{clustered about median} \end{cases}.$$

To give you an idea of what this means, the "curve" that you have heard so much, the one that students keep asking *Are you going to grade on the "curve?"*, is clustered about the median.

If $\mathbb{C} \geq 0.4$ the sample cannot have any outliers. If neither Q_1 nor Q_3 is a sample extreme then it is very unlikely for there to be an outlier when $\mathbb{C} \geq 0.25$.

Example 1:
Determine whether the two-success data are clustered about the mean, dispersed from the mean, or neither.

Answer:
The quartiles from the median-of-medians, traditional sample and restricted proper methods were calculated in Example 5, §5.10, starting on page 262, and are summarized below.

Method	Q_1	Q_3
median-of-medians	4.0	7.5
traditional sample	3.5	9.0
restricted proper	$\frac{231}{62}$	$\frac{53}{6}$

In addition, $x_{\min} = 2$ and $x_{\max} = 28$, so $R = 28 - 2 = 26$.

By the median-of-medians method,

$$R_Q = 7.5 - 4 = 3.5; \text{ so, } \mathbb{C} = \frac{3.5}{26} = \frac{7}{52} \approx 0.1346 \leq 0.3.$$

We conclude the data are clustered about the median.

By the traditional sample method,

$$R_Q = 9 - 3.5 = 5.5; \text{ so, } \mathbb{C} = \frac{5.5}{26} = \frac{11}{52} \approx 0.2115 \leq 0.3.$$

We conclude the data are clustered about the median.

By the restricted proper method,

$$R_Q = \frac{53}{6} - \frac{231}{62} = \frac{475}{93}; \text{ so, } \mathbb{C} = \frac{475/93}{26} = \frac{475}{2418} \approx 0.1964 \leq 0.3.$$

We conclude that the data are clustered about the median.

6.1. RANGE & INTERQUARTILE RANGE

Note: The midrange of the data is

$$x_{\text{mid}} = \frac{2+28}{2} = 15,$$

which is well above even the highest estimate for Q_3. It would be extremely unusual for the data *not* to be clustered in this circumstance.

Example 2:
Determine the clustering factor of the random gamma data. What do you conclude from this?

Answer:
The quartiles for this data were determined in Example 6, §5.10, which started on page 265. The results are summarized below.

Method	Q_1	Q_3
median-of-medians	9	17
traditional sample	9	18
restricted proper	$\frac{26}{3}$	18

Furthermore, $x_{\min} = 4$ and $x_{\max} = 31$; thence, $R = 31 - 4 = 27$.
By the median-of-medians method,

$$R_Q = 17 - 9 = 8 \text{ and } \mathbb{C} = \frac{8}{27} \approx 0.2963 \leq 0.3.$$

From this, the data are clustered about the median.
By the traditional sample method,

$$R_Q = 18 - 9 = 9 \text{ and } \mathbb{C} = \frac{9}{27} = \frac{1}{3} \approx 0.3333 > 0.3.$$

As $0.3333 < 0.7$, we conclude the data are neither clustered nor disperse.
By the restricted proper method,

$$R_Q = 18 - \frac{26}{3} = \frac{28}{3} \text{ and } \mathbb{C} = \frac{28/3}{27} = \frac{28}{81} \approx 0.3457 > 0.3.$$

As $0.3457 < 0.7$, we determine that the data are neither clustered nor disperse.

If the median is a repeated observation, then R_Q is always smaller from the median-of-medians method than from the traditional sample method. For this reason it is rarely used to calculate the quartiles when determining $\mathbb{C}$. The value of $\mathbb{C}$ is small enough that this difference in R_Q just tipped the edge of 0.3 for the median-of-medians method and gave us a different result.

Example 3:
Determine whether the oyster data are clustered about the mean.

The sorted data are

5.76	6.10	6.36	6.37	6.42	6.53	6.63	6.67	6.74	6.91
6.91	6.97	7.03	7.04	7.37	7.40	7.71	7.97	8.02	8.04
8.12	8.14	8.42	8.53	8.61	8.62	9.01	9.20	9.21	9.39
9.47	9.53	9.55	9.98	10.24	10.50	10.62	10.63	10.63	10.81
10.81	10.86	11.02	11.09	11.17	11.31	11.59	11.97	12.07	13.18

There are 50 data with $x_{\min} = 5.76$ and $x_{\max} = 13.18$; whereby, $R = 13.18 - 5.76 = 7.42$. We note that both extremes are unique.
By the median-of-medians method,

$$\#(M) = \frac{1+50}{2} = 25.5 \text{ and } M = \frac{x_{(25)} + x_{(26)}}{2} = \frac{8.61 + 8.62}{2} = 8.615.$$

The largest k such that $x_{(k)} \leq M$ is $k = 25$ as $x_{(25)} = 8.61 < 8.615$ and $x_{(26)} = 8.62 > 8.615$. This also means that the smallest j such that $x_{(j)} \geq M$ is $j = 26$. We conclude,

$$\#(Q_1) = \frac{1+25}{2} = 13 \text{ and } \#(Q_3) = \frac{26+50}{2} = 38.$$

6.1. RANGE & INTERQUARTILE RANGE

Thus, $Q_1 = x_{(13)} = 7.03$ and $Q_3 = x_{(38)} = 10.63$. The interquartile range is $R_Q = 10.63 - 7.03 = 3.6$. We conclude,

$$\mathbb{C} = \frac{3.6}{7.42} = \frac{180}{371} \approx 0.4852.$$

By the traditional sample method

$$\#(Q_1) = \frac{50}{4} = 12.5 \text{ and } \#(Q_3) = 3 \cdot 12.5 = 37.5.$$

Therefore, the quartiles agree with the median-of-medians method.
By the restricted proper method,

$$\#(Q_1) = \frac{3 \cdot 1 + 50}{4} = 13.25 \text{ and } \#(Q_3) = \frac{1 + 3 \cdot 50}{4} = 37.75.$$

In the quest for Q_1, $x_{(13)} = 7.03$ is unique, so $y_1 = 7.03$, $m_1 = 13$, $y_2 = 7.04$ and $m_2 = 14$. We conclude

$$Q_1 = \frac{7.03 \cdot (14 - 13.25) + 7.04 \cdot (13.25 - 13)}{14 - 13} = 5.2725 + 1.76 = 7.0325.$$

Going for Q_3 we observe $x_{(37)} = 10.62$ is unique, so $y_1 = 10.62$, $m_1 = 37$, $y_2 = 10.63$ and $m_2 = (38 + 39)/2 = 38.5$. By interpolation,

$$Q_3 = \frac{10.62 \cdot (38.5 - 37.75) + 10.63 \cdot (37.75 - 37)}{38.5 - 37}$$
$$= \frac{7.965 + 7.9725}{1.5} = \frac{85}{8} = 10.625.$$

From the above,

$$R_Q = 10.625 - 7.0325 = 3.5925 \text{ and } \mathbb{C} = \frac{3.5925}{7.42} = \frac{1437}{2968} \approx 0.4842.$$

In all cases we conclude the data are not clustered about the median. [They are not dispersed from the median either, but that was not asked.]

Variation Statistics

THERE ARE SEVERAL estimates for the average deviation based upon the range and interquartile range that we will now look at.

The **average variation** is the average distance between consecutive sorted observations. This is denoted v — a minuscule v (the majuscule V is used for a different statistic). Some people use upsilon, υ, or nu, ν, (which look pretty similar, especially the upsilon) for this statistic. The average variation is defined as

$$v = \frac{R}{n-1}.$$

Suppose there are k distinct observations denoted $y_1, \ldots, y_k$ — where we assume these are sorted from smallest to largest. When discussing grouping, we defined the granularity, G. We recall that definition here. Let $w_i = y_{i+1} - y_i$ for $i = 1, \ldots, k-1$. The **granularity** was defined as

$$G = \max(2\min_i(w_i), \max_i(w_i));$$

that is, it is the larger of twice the minimum consecutive difference and the maximum consecutive difference. A similar statistic is the grain. The **grain** is the average difference between consecutive distinct observations in the sorted order. It is denoted **gr**. Thus,

$$\text{gr} = \frac{R}{k-1}.$$

The **central error**, denoted c_{err}, is an estimate of the average deviation from the median. Its definition depends upon the clustering

6.1. RANGE & INTERQUARTILE RANGE

factor.

$$c_{\text{err}} = \begin{cases} \mathbb{C} \leq 0.5 & \dfrac{R}{6} \\ \mathbb{C} > 0.5 & \dfrac{R}{4} \end{cases}.$$

Another statistic to estimate the average deviation from the median is the **central deviance**, denoted c_{dev}. This is defined as

$$c_{\text{dev}} = \frac{20}{27} \cdot R_Q.$$

Example 4:
Determine the average variation, central error and central deviance for the call data introduced in Example 3, §3.4, starting on page 69.
Answer:
The sorted data are given below.

24	24	24	24	26	26	27	27	29	29
29	30	30	31	32	33	35	35	36	37
37	38	38	39	40	42	42	43	43	44
45	45	46	47	48	48	49	50	59	66
66	71	72	75	76	79	84	88	91	112

From the above,

$$R = 112 - 24 = 88; \text{ so, } v = \frac{88}{50-1} = \frac{88}{49} \approx 1.7959.$$

We need the interquartile range for the rest of this. By the median-of-medians method,

$$\#(M) = \frac{1+50}{2} = 25.5 \text{ and } M = \frac{x_{(25)} + x_{(26)}}{2} = \frac{40+42}{2} = 41.$$

The highest integer k such that $x_{(k)} \leq M$ is $k = 25$ and the smallest integer j such that $x_{(j)} \geq M$ is $j = 26$. Therefore,

$$\#(Q_1) = \frac{1+25}{2} = 13 \text{ and } \#(Q_3) = \frac{26+50}{2} = 38.$$

From this we get $Q_1 = x_{(13)} = 30$ and $Q_3 = x_{(38)} = 50$. The interquartile range is
$$R_Q = 50 - 30 = 20.$$
This makes
$$\mathbb{C} = \frac{20}{88} = \frac{5}{22} \approx 0.2273 \leq 0.5; \text{ so, } c_{\text{err}} = \frac{88}{6} = \frac{44}{3} \approx 14.6667.$$
Furthermore, the central deviance is
$$c_{\text{dev}} = \frac{20 \cdot 20}{27} = \frac{400}{27} \approx 14.8148.$$
The traditional sample method makes
$$\#(Q_1) = \frac{50}{4} = 12.5 \text{ and } \#(Q_3) = 3 \cdot 12.5 = 37.5.$$
This makes the quartiles the same as given by the median-of-medians method.

The restricted proper method notes that
$$\#(x_{\min}) = \frac{1+4}{2} = 2.5 \text{ and } \#(x_{\max}) = 50.$$
We conclude
$$\#(Q_1) = \frac{3 \cdot 2.5 + 50}{4} = \frac{115}{8} = 14.375$$
and
$$\#(Q_3) = \frac{2.5 + 3 \cdot 50}{4} = \frac{305}{8} = 38.125.$$
Both $x_{(14)} = 31$ and $x_{(15)} = 32$ are unique, so
$$Q_1 = \frac{31 \cdot (15 - 14.375) + 32 \cdot (14.375 - 14)}{15 - 14} = 19.375 + 12 = 31.375.$$
Both $x_{(38)} = 50$ and $x_{(39)} = 59$ are also unique; hence,
$$Q_3 = \frac{50 \cdot (39 - 38.125) + 59 \cdot (38.125 - 38)}{39 - 38} = 43.75 + 7.375 = 51.125.$$

6.1. RANGE & INTERQUARTILE RANGE

We conclude $R_Q = 51.125 - 31.375 = 19.75$. Thence,

$$\mathbb{C} = \frac{19.75}{88} = \frac{79}{352} \approx 0.2244 \leq 0.5; \text{ so, } c_{\text{err}} = \frac{88}{6} = \frac{44}{3} \approx 14.6667.$$

Furthermore, the central deviance is

$$c_{\text{dev}} = \frac{20 \cdot 19.75}{27} = \frac{395}{27} \approx 14.6296.$$

Example 5:
Compare the grain to the granularity for the call data.

Answer:
The frequency table for the data is

24: 4	31: 1	37: 2	43: 2	48: 2	71: 1	84: 1
26: 2	32: 1	38: 2	44: 1	49: 1	72: 1	88: 1
27: 2	33: 1	39: 1	45: 2	50: 1	75: 1	91: 1
29: 3	35: 2	40: 1	46: 1	59: 1	76: 1	112: 1
30: 2	36: 1	42: 2	47: 1	66: 2	79: 1	

There are 34 distinct observations; whence, the grain is

$$\text{gr} = \frac{88}{34 - 1} = \frac{8}{3} \approx 2.6667.$$

The table of consecutive differences is

2	1	2	1	1	1	1	2	1	1	1
1	1	2	1	1	1	1	1	1	1	
9	7	5	1	3	1	3	5	4	3	21

The smallest difference is 1, which first occurs as w_2. The largest difference is $w_{33} = 21$. This makes the granularity

$$G = \max(2 \cdot 1, 21) = 21.$$

That 21 is very large in comparison to the other differences. Because this is the final difference, 112 is almost certainly an outlier. If we eliminate that point, the next largest difference is $9 > 2$, so the granularity becomes $G = 9$. It is this granularity that should probably be used if we were considering grouping. We leave it to you to determine the proper number of groups needed to get up to 91.

Example 6:
Determine the average variation, central error, central deviance and grain for the dice roll data.
Answer:
The sorted data are given below.

$$
\begin{array}{cccccccccc}
2 & 3 & 3 & 3 & 3 & 3 & 4 & 4 & 4 & 4 \\
5 & 5 & 5 & 5 & 5 & 5 & 6 & 6 & 6 & 6 \\
6 & 6 & 6 & 7 & 7 & 7 & 7 & 7 & 7 & 7 \\
7 & 8 & 8 & 8 & 8 & 8 & 8 & 8 & 9 & 9 \\
9 & 9 & 9 & 10 & 10 & 10 & 10 & 10 & 11 & 12
\end{array}
$$

From the above, the range and average variation are

$$R = 12 - 2 = 10; \text{ thence, } v = \frac{10}{50-1} = \frac{10}{49} \approx 0.2041.$$

There are 11 distinct observations, so the grain is

$$\text{gr} = \frac{10}{11-1} = 1.$$

Although not asked about, all differences are 1, so $G = 2$.

The rest of the answers require the interquartile range There are 50 data of which the extremes are both unique. We note that both are also population extremes.

By the median-of-medians method, $\#(M) = (1+50)/2 = 25.5$. Now, $x_{(25)} = x_{(26)} = 7$, so $M = 7$. The highest integer k such that $x_{(k)} \leq M$ is $k = 31$. The lowest integer j such that $x_{(j)} \geq M$ is $j = 24$. We conclude,

$$\#(Q_1) = \frac{1+31}{2} = 16 \text{ and } \#(Q_3) = \frac{24+50}{2} = 37.$$

6.1. RANGE & INTERQUARTILE RANGE

Thus, $Q_1 = x_{(16)} = 5$ and $Q_3 = x_{(37)} = 8$. The interquartile range is therefore,
$$R_Q = 8 - 5 = 3 \text{ and } \mathbb{C} = \frac{3}{10} = 0.3 \leq 0.5.$$
We therefore have cerr $= R/6 = 5/3 \approx 1.6667$. The central deviance is
$$c_{\text{dev}} = \frac{20 \cdot 3}{27} = \frac{20}{9} \approx 2.2222.$$

By the traditional sample method,
$$\#(Q_1) = \frac{50}{4} = 12.5 \text{ and } \#(Q_3) = 3\#(Q_1) = 37.5.$$

Wherefrom, $Q_1 = x_{(13)} = 5$ and $Q_3 = x_{(38)} = 8$. These have the same values as the median-of-medians method, but only because there is much repetition.

By the restricted proper method,
$$\#(Q_1) = \frac{3 \cdot 1 + 50}{4} = 13.25 \text{ and } \#(Q_3) = \frac{1 + 3 \cdot 50}{4} = 37.75.$$

For Q_1, $x_{(13)} = 5$ and $\#(5) = (11+16)/2 = 13.5 > 13.25$. Whereby, $y_2 = 5$, $m_2 = 13.5$, $y_1 = 4$ and $m_1 = (7+10)/2 = 8.5$. By interpolation,
$$Q_1 = \frac{4 \cdot (13.5 - 13.25) + 5 \cdot (13.25 - 8.5)}{13.5 - 8.5} = \frac{1 + 23.75}{5} = \frac{99}{20} = 4.95.$$

For Q_3, $x_{(37)} = 8$ and $\#(8) = (32+38)/2 = 35 < 37.75$. Thereby, $y_1 = 8$, $m_1 = 35$, $y_2 = 9$ and $m_2 = (39+43)/2 = 41$. By interpolation,
$$Q_3 = \frac{8 \cdot (41 - 37.75) + 9 \cdot (37.75 - 35)}{41 - 35} = \frac{26 + 24.75}{6} = \frac{203}{24} \approx 8.4583.$$

The interquartile range is
$$R_Q = \frac{203}{24} - \frac{99}{20} = \frac{421}{120} \approx 3.5083.$$

We conclude
$$\mathbb{C} = \frac{421/120}{10} = \frac{421}{1200} \approx 0.3508 \leq 0.5$$
and $c_{\text{err}} = 10/6 = 5/3 \approx 1.6667$. Furthermore, the central deviance is
$$c_{\text{dev}} = \frac{20}{27} \cdot \frac{421}{120} = \frac{421}{162} \approx 2.5988.$$

Exercises 6.1

1. One day while walking through a field you notice there seem to be a lot of stones lying on the ground. Curious about what the average size of a stone is in the field you randomly collect 30 stones. You weigh each stone and record its weight in grams. The following table shows the raw data.

110	126	75	110	126	94
90	112	111	139	131	92
102	103	96	75	67	101
104	111	100	65	92	93
106	104	93	115	109	95

    Determine the average variation, central error, central deviance, grain and granularity.

2. The scores, in the order graded, on a recent midterm examination in Statistics, a class with 30 students, are given below.

70	80	99	98	85	89
87	79	83	38	69	70
60	69	78	40	75	56
70	51	99	69	95	86
57	53	47	50	55	81

Determine the average variation, central error, central deviance, grain and granularity.

3. Below is a sample of the monthly salary for 30 randomly selected people in mid-management in thousands of dollars.

3.08	3.18	3.13	7.41	3.73	3.05
4.17	3.58	3.36	3.27	3.36	4.74
3.32	3.61	4.26	3.02	3.45	4.06
6.73	4.72	3.13	3.15	3.70	3.59
3.12	3.61	5.03	3.53	3.20	3.32

Determine the average variation, central error, central deviance, grain and granularity.

4. A pair of dice, one red and one green, is rolled in the following manner. On the first toss both dice are tossed. Then the red die is tossed while the green die keeps its value to get the second sum. Then the green die is tossed while the red die keeps its value. You continue tossing alternately the red then the green, die and generate the following table.

8	11	8	6	10	9	9
7	2	5	10	11	8	5
3	6	9	7	7	9	10
7	3	3	7	9	8	7
5	4	8	11	11	10	10

Determine the average variation, central error, central deviance, grain and granularity.

5. The following random sample of 72 observations comes from a gamma distribution, so $\mu_{min} = 0$ and μ_{max} has no upper limit. The raw data are

2.15	1.92	1.92	2.20	3.67	4.55	3.32	2.71	4.87
2.09	1.13	1.09	2.12	4.89	3.03	3.17	4.74	4.44
4.61	2.78	2.97	2.78	2.88	3.66	2.55	4.40	3.54
2.18	3.43	3.84	4.60	1.76	2.85	2.27	1.38	1.24
4.22	1.67	4.51	3.50	6.69	5.92	6.42	0.95	4.57
3.10	3.26	2.95	2.45	1.95	2.06	3.10	2.03	4.98
0.86	2.61	1.92	2.78	1.01	2.64	3.14	0.86	3.68
2.71	3.06	4.33	4.63	2.47	2.92	1.38	2.81	5.71

Determine the average variation, central error, central deviance, grain and granularity.

6.2 Average Absolute Variability

T<small>HE AVERAGE VARIATION</small> calculated in the last section assumes that all observations are distinct and is actually the average separation between consecutive values under that assumption. What you really want to know is the average separation between two randomly selected observations. There are a number of ways of estimating this and in this section we consider one of them.

The **average absolute variability**, denoted **aav**, is the arithmetic average of all differences $x_{(j)} - x_{(i)}$ such that $1 \leq i < j \leq n$. There are
$$\binom{n}{2} = \frac{n \cdot (n-1)}{2}$$
such differences. Thus,
$$\text{aav} = \frac{2 \sum_{i=1}^{n-1} \sum_{j=i+1}^{n} (x_{(j)} - x_{(i)})}{n \cdot (n-1)}.$$

6.2. AVERAGE ABSOLUTE VARIABILITY

Suppose there are k distinct observations $y_1, \ldots, y_k$ — sorted from smallest to largest — and that $\text{frq}(y_i) = f_i$, for $i = 1, \ldots, k$. Then there are
$$\binom{f_i}{2} = \frac{f_i \cdot (f_i - 1)}{2}$$
differences between pairs having the value y_i and
$$f_i \cdot f_j$$
differences between pairs having distinct values. The above formula becomes
$$\text{aav} = \frac{2 \sum_{i=1}^{k-1} \sum_{j=i+1}^{k} (f_i \cdot f_j) \cdot (y_{(j)} - y_{(i)})}{n \cdot (n-1)}.$$
Although the formula does not calculate the zero differences, we do not ignore them. This can be seen from the denominator.

Example 1:
Determine the average absolute variability of the two-success data.
Answer:
The sorted data is

2	2	2	2	2	2	2	2	2	3
3	3	3	3	3	3	3	3	3	3
3	3	3	3	3	4	4	4	4	4
4	4	4	4	4	4	4	4	4	4
5	5	5	5	5	5	5	5	5	5
5	5	5	5	5	5	5	6	6	6
6	6	6	6	6	6	7	7	7	7
8	8	8	8	9	9	9	9	9	10
10	10	10	10	10	11	11	11	11	12
12	14	14	14	14	14	15	16	17	28

292   CHAPTER 6. DATA VARIABILITY

This gives the following tallies

   2: 9   4: 15   6: 9   8: 4   10: 6   12: 2   15: 1   17: 1
   3: 16   5: 17   7: 4   9: 5   11: 4   14: 5   16: 1   28: 1

We again create a table of differences for the different distinct values. This produces the following absolute difference table.

	3	4	5	6	7	8	9	10	11	12	14	15	16	17	28
2	1 144	2 135	3 153	4 81	5 36	6 36	7 45	8 54	9 36	10 18	12 45	13 9	14 9	15 9	26 9
3		1 240	2 272	3 144	4 64	5 64	6 80	7 96	8 64	9 32	11 80	12 16	13 16	14 16	25 16
4			1 255	2 135	3 60	4 60	5 75	6 90	7 60	8 30	10 75	11 15	12 15	13 15	24 15
5				1 153	2 68	3 68	4 85	5 102	6 68	7 34	9 85	10 17	11 17	12 17	23 17
6					1 36	2 36	3 45	4 54	5 36	6 18	8 45	9 9	10 9	11 9	22 9
7						1 16	2 20	3 24	4 16	5 8	7 20	8 4	9 4	10 4	21 4
8							1 20	2 24	3 16	4 8	6 20	7 4	8 4	9 4	20 4
9								1 30	2 20	3 10	5 25	6 5	7 5	8 5	19 5
10									1 24	2 12	4<>30	5 6	6 6	7 6	18 6
11										1 8	3 20	4 4	5 4	6 4	17 4
12											2 10	3 2	4 2	5 2	16 2
14												1 5	2 5	3 5	14 5
15													1 1	2 1	13 1
16														1 1	12 1
17															11 1

Putting in the zero differences and combining differences of the same magnitude gives the following frequency table of absolute differences be-

6.2. AVERAGE ABSOLUTE VARIABILITY

tween all pairs of observations

0: 487	3: 547	6: 327	9: 170	12: 94	15: 9	18: 6	21: 4	24: 15
1: 933	4: 404	7: 270	10: 123	13: 41	16: 2	19: 5	22: 9	25: 16
2: 738	5: 358	8: 206	11: 122	14: 30	17: 4	20: 4	23: 17	26: 9

The average absolute variability is

$$\text{aav} = \frac{1 \cdot 933 + \cdots + 26 \cdot 9}{4950} = \frac{21324}{4950} = \frac{3554}{825} \approx 4.3079.$$

In the previous example there should be $100 \cdot 99/2 = 4950$ different absolute differences and we leave it to you to verify that is the sum of the frequencies in the frequency table of absolute differences.

Example 2:
Determine the average absolute variability for the dice roll data.

Answer:
The sorted data are

2	3	3	3	3	3	4	4	4	4
5	5	5	5	5	5	6	6	6	6
6	6	6	7	7	7	7	7	7	7
7	8	8	8	8	8	8	8	9	9
9	9	9	10	10	10	10	10	11	12

The tally table from the above is

2: 1	4: 4	6: 7	8: 7	10: 5	12: 1
3: 5	5: 6	7: 8	9: 5	11: 1	

The absolute differences between different observed values are given in the following table — we have allowed no space for the zero differences or their frequencies in this table.

	3	4	5	6	7	8	9	10	11	12
2	1	2	3	4	5	6	7	8	9	10
	5	4	6	7	8	7	5	5	1	1
3		1	2	3	4	5	6	7	8	9
		20	30	35	40	35	25	25	5	5
4			1	2	3	4	5	6	7	8
			24	28	32	28	20	20	4	4
5				1	2	3	4	5	6	7
				42	48	42	30	30	6	6
6					1	2	3	4	5	6
					56	49	35	35	7	7
7						1	2	3	4	5
						56	40	40	8	8
8							1	2	3	4
							35	35	7	7
9								1	2	3
								25	5	5
10									1	2
									5	5
11										1
										1

This gives us the following tally table of differences

0: 121   2: 244   4: 155   6: 65   8: 14   10: 1
1: 268   3: 202   5: 108   7: 41   9: 6

The average absolute variability is (where we again ignore the zeros except in the denominator total)

$$\text{aav} = \frac{1 \cdot 268 + \cdots + 10 \cdot 1}{1225} = \frac{3375}{1225} = \frac{135}{49} \approx 2.7551$$

Grouped Data

IT IS POSSIBLE to estimate the average absolute variability with grouped data using the midpoints of the intervals as observations and the number of observations in each interval as the frequency the

6.2. AVERAGE ABSOLUTE VARIABILITY

midpoint. Empty intervals are ignored, just as they were with Walsh sums.

Example 3:
Determine estimates for the average absolute variability of the random gamma data using four intervals. Repeat the estimate with five intervals.

Answer:
For four intervals, the common width is

$$w = \frac{31 - 4}{4} = 6.75.$$

We will round this up to the nearest 0.1 and make $w = 6.8$. The starting point for the intervals is

$$t_0 = \frac{4 + 31 - 4 \cdot 6.8}{2} = 3.9.$$

This gives the following intervals, midpoints, and frequencies.

I_i	[3.9, 10.7)	[10.7, 17.5)	[17.5, 24.3)	[24.3, 31.1]
m_i	7.3	14.1	20.9	27.7
f_i	9	9	5	2

The above gives the following difference table.

	14.1	20.9	27.7
7.3	6.8	13.6	20.4
	81	45	18
14.1		6.8	13.6
		45	18
20.9			6.8
			10

The above produces the following frequencies of differences

0.0: 83   6.8: 136   13.6: 63   20.4: 18

$83 + 136 + 63 + 18 = 300 = 25 \cdot 24/2$, as it should. This means
$$\text{aav} = \frac{6.8 \cdot 136 + 13.6 \cdot 63 + 20.9 \cdot 18}{300} = \frac{2148.8}{300} = \frac{2686}{375} \approx 7.1627.$$
With five intervals
$$w = \frac{31 - 4}{5} = 5.4.$$
This gives the following intervals, midpoints and frequencies.

I_i	$[4, 9.4)$	$[9.4, 14.8)$	$[14.8, 20.2)$	$[20.2, 25.6)$	$[25.6, 31]$
m_i	6.7	12.1	17.5	22.9	28.3
f_i	8	7	4	4	2

The difference table for this data is

	12.1	17.5	22.5	28.3
6.7	5.4	10.8	16.2	21.6
	56	32	32	16
12.1		5.4	10.8	16.2
		28	28	14
17.5			5.4	10.8
			16	8
22.9				5.4
				8

The above table gives the following tallies of differences.

$$0: 62 \quad 5.4: 108 \quad 10.8: 68 \quad 16.2: 46 \quad 21.6: 16$$

Again these sum to 300 as they should. We conclude
$$\text{aav} = \frac{5.4 \cdot 108 + \cdots + 21.6 \cdot 16}{300} = \frac{2408.4}{300} = \frac{2007}{250} = 8.028.$$

6.2. AVERAGE ABSOLUTE VARIABILITY

Exercises 6.2

1. One day while walking through a field you notice there seem to be a lot of stones lying on the ground. Curious about what the average size of a stone is in the field you randomly collect 30 stones. You weigh each stone and record its weight in grams. The following table shows the raw data.

110	126	75	110	126	94
90	112	111	139	131	92
102	103	96	75	67	101
104	111	100	65	92	93
106	104	93	115	109	95

    Determine the average absolute variability for this data. If there are more than 16 distinct data values, then use the grouped method.

2. The scores, in the order graded, on a recent midterm examination in Statistics, a class with 30 students, are given below.

70	80	99	98	85	89
87	79	83	38	69	70
60	69	78	40	75	56
70	51	99	69	95	86
57	53	47	50	55	81

    Determine the average absolute variability for this data. If there are more than 16 distinct data values, then use the grouped method.

3. Below is a sample of the monthly salary for 30 randomly selected people in mid-management in thousands of dollars.

3.08	3.18	3.13	7.41	3.73	3.05
4.17	3.58	3.36	3.27	3.36	4.74
3.32	3.61	4.26	3.02	3.45	4.06
6.73	4.72	3.13	3.15	3.70	3.59
3.12	3.61	5.03	3.53	3.20	3.32

Determine the average absolute variability for this data. If there are more than 16 distinct data values, then use the grouped method.

4. A pair of dice, one red and one green, is rolled in the following manner. On the first toss both dice are tossed. Then the red die is tossed while the green die keeps its value to get the second sum. Then the green die is tossed while the red die keeps its value. You continue tossing alternately the red then the green, die and generate the following table.

8	11	8	6	10	9	9
7	2	5	10	11	8	5
3	6	9	7	7	9	10
7	3	3	7	9	8	7
5	4	8	11	11	10	10

Determine the average absolute variability for this data. If there are more than 16 distinct data values, then use the grouped method.

5. The following random sample of 72 observations comes from a gamma distribution, so $\mu_{min} = 0$ and μ_{max} has no upper limit. The raw data are

2.15	1.92	1.92	2.20	3.67	4.55	3.32	2.71	4.87
2.09	1.13	1.09	2.12	4.89	3.03	3.17	4.74	4.44
4.61	2.78	2.97	2.78	2.88	3.66	2.55	4.40	3.54
2.18	3.43	3.84	4.60	1.76	2.85	2.27	1.38	1.24
4.22	1.67	4.51	3.50	6.69	5.92	6.42	0.95	4.57
3.10	3.26	2.95	2.45	1.95	2.06	3.10	2.03	4.98
0.86	2.61	1.92	2.78	1.01	2.64	3.14	0.86	3.68
2.71	3.06	4.33	4.63	2.47	2.92	1.38	2.81	5.71

Determine the average absolute variability for this data. If there are more than 16 distinct data values, then use the grouped method.

6.3 Median Absolute Variability

THE PROBLEM WITH the average absolute variability is that a single outlier can overwhelm the average and make the data appear more variable than it actually is. A better estimate is the **median absolute variability**, denoted **mav**. As its name implies, the **median absolute variability** is the median of the differences between observations.

Example 1:
Determine the median absolute variability for the dice roll data.

Answer:
In Example 2, §6.2, starting on page 293, we determined all differences and their frequencies. The tally table for the differences is repeated here.

| 0: 121 | 2: 244 | 4: 155 | 6: 65 | 8: 14 | 10: 1 |
| 1: 268 | 3: 202 | 5: 108 | 7: 41 | 9: 6 | |

There are 1225 differences.

By the median-of-medians method,

$$\#(M) = \frac{1 + 1225}{2} = 613.$$

Therefore, the median is

$$M = d_{(613)} = 2.$$

The traditional sample method calculates $\#(M) = 1225/2 = 612.5$, but gets the identical result.

In the restricted proper method,

$$\#(d_{\min}) = \frac{1 + 121}{2} = 61 \text{ and } \#(d_{\max}) = 1225.$$

Thus,
$$\#(M) = \frac{61 + 1225}{2} = 643.$$
For this, $d_{(643)} = 3$ and $\#(3) = (634+835)/2 = 734.5 > 643$. Whence, $y_2 = 3$, $m_2 = 734.5$, $y_1 = 2$ and $m_1 = (390 + 633)/2 = 511.5$. By interpolation,
$$M = \frac{2 \cdot (734.5 - 643) + 3 \cdot (643 - 511.5)}{734.5 - 511.5}$$
$$= \frac{183 + 394.5}{223} = \frac{1155}{446} \approx 2.5897.$$

Example 2:
Determine the median absolute variability of the two-success data.

Answer:
 In Example 1, §6.2, starting on page 291, we determined all differences and their frequencies. The tally table for the differences is repeated here.

0: 487	3: 547	6: 327	9: 170	12: 94	15: 9	18: 6	21: 4	24: 15
1: 933	4: 404	7: 270	10: 123	13: 41	16: 2	19: 5	22: 9	25: 16
2: 738	5: 358	8: 206	11: 122	14: 30	17: 4	20: 4	23: 17	26: 9

There are 4950 differences.
 By the median-of-medians method,
$$\#(M) = \frac{1 + 4950}{2} = 2475.5.$$
Observe that $d_{(2475)} = d_{(2476)} = 3$, so $M = 3$.
 The traditional sample method calculates
$$\#(M) = \frac{4950}{2} = 2475,$$
but it gives results identical to those for the median-of-medians method.

6.3. MEDIAN ABSOLUTE VARIABILITY

The restricted proper method begins by determining the effective addresses for the sample extremes.

$$\#(d_{\min}) = \frac{1 + 487}{2} = 244 \text{ and } \#(d_{\max}) = \frac{4942 + 4950}{2} = 4946.$$

This means

$$\#(M) = \frac{244 + 4946}{2} = 2595.$$

For M, $d_{(2595)} = 3$ and $\#(3) = (2159 + 2705)/2 = 2432 < 2595$. Whereby, $y_1 = 3$, $m_1 = 2432$, $y_2 = 4$ and $m_2 = (2706 + 3109)/2 = 2907.5$. By interpolation,

$$M = \frac{3 \cdot (2907.5 - 2595) + 4 \cdot (2595 - 2432)}{2907.5 - 2432}$$
$$= \frac{937.5 + 652}{475.5} = \frac{3179}{951} \approx 3.3428.$$

Grouped Data

WITH GROUPED data, the midpoints of the intervals become the observations and the numbers of observations in each interval become the frequencies for those midpoints.

Example 3:
Determine estimates for the median absolute variability of the random gamma data using four intervals. Repeat the estimate with five intervals.

Answer:
The tally table for the differences with four intervals, taken from Example 3, §6.2, starting on page 295, is

0.0: 83   6.8: 136   13.6: 63   20.4: 18

There are 300 differences in total.

By the median-of-medians method, $\#(M) = (1+300)/2 = 150.5$, while by the traditional sample method $\#(M) = 300/2 = 150$. In either event, the median is the same for these methods, namely

$$M = \frac{d_{(150)} + d_{(151)}}{2} = \frac{6.8 + 6.8}{2} = 6.8.$$

The effective addresses of the sample extremes are

$$\#(d_{\min}) = \frac{1+83}{2} = 42 \text{ and } \#(d_{\max}) = \frac{283+300}{2} = 291.5.$$

The median has effective address

$$\#(M) = \frac{42 + 291.5}{2} = 166.75.$$

For this, $d_{(166)} = 6.8$ and $\#(6.8) = (84+219)/2 = 151.5 < 166.75$. Thence, $y_1 = 6.8$, $m_1 = 151.5$, $y_2 = 13.6$ and $m_2 = (220+282)/2 = 251$. By interpolation,

$$M = \frac{6.8 \cdot (251 - 166.75) + 13.6 \cdot (166.75 - 151.5)}{251 - 151.5}$$

$$= \frac{572.9 + 207.4}{99.5} = \frac{7803}{995} \approx 7.8422.$$

When there are five intervals, the tally table of differences, taken from Example 3, §6.2, starting on page 295, is

0: 62   5.4: 108   10.8: 68   16.2: 46   21.6: 16

The sum of the frequencies is still 300 as there were 25 observations in the original data set and $25 \cdot 24/2 = 300$.

By the median-of-medians method, $\#(M) = (1+300)/2 = 150.5$, while by the traditional sample method $\#(M) = 300/2 = 150$. In either event, the median is the same for these methods, namely

$$M = \frac{d_{(150)} + d_{(151)}}{2} = \frac{5.4 + 5.4}{2} = 5.4.$$

6.3. MEDIAN ABSOLUTE VARIABILITY

The effective addresses of the sample extremes are
$$\#(d_{\min}) = \frac{1+62}{2} = 31.5 \text{ and } \#(d_{\max}) = (285+300)/2 = 292.5.$$
From this, the restricted proper method calculates
$$\#(M) = \frac{31.5 + 292.5}{2} = 162.$$
Now, $d_{(162)} = 5.4$ and $\#(5.4) = (63+170)/2 = 116.5 < 162$. Wherefrom, $y_1 = 5.4$, $m_1 = 116.5$, $y_2 = 10.8$ and $m_2 = (171+238)/2 = 204.5$. By interpolation,
$$M = \frac{5.4 \cdot (204.5 - 162) + 10.8 \cdot (162 - 116.5)}{204.5 - 116.5}$$
$$= \frac{229.5 + 491.4}{88} = \frac{7209}{880} \approx 8.1920.$$

Exercises 6.3

1. One day while walking through a field you notice there seem to be a lot of stones lying on the ground. Curious about what the average size of a stone is in the field you randomly collect 30 stones. You weigh each stone and record its weight in grams. The following table shows the raw data.

110	126	75	110	126	94
90	112	111	139	131	92
102	103	96	75	67	101
104	111	100	65	92	93
106	104	93	115	109	95

   Determine the median absolute variability for this data. If there are more than 16 distinct data values, then use the grouped method.

2. The scores, in the order graded, on a recent midterm examination in Statistics, a class with 30 students, are given below.

$$\begin{array}{cccccc} 70 & 80 & 99 & 98 & 85 & 89 \\ 87 & 79 & 83 & 38 & 69 & 70 \\ 60 & 69 & 78 & 40 & 75 & 56 \\ 70 & 51 & 99 & 69 & 95 & 86 \\ 57 & 53 & 47 & 50 & 55 & 81 \end{array}$$

Determine the median absolute variability for this data. If there are more than 16 distinct data values, then use the grouped method.

3. Below is a sample of the monthly salary for 30 randomly selected people in mid-management in thousands of dollars.

$$\begin{array}{cccccc} 3.08 & 3.18 & 3.13 & 7.41 & 3.73 & 3.05 \\ 4.17 & 3.58 & 3.36 & 3.27 & 3.36 & 4.74 \\ 3.32 & 3.61 & 4.26 & 3.02 & 3.45 & 4.06 \\ 6.73 & 4.72 & 3.13 & 3.15 & 3.70 & 3.59 \\ 3.12 & 3.61 & 5.03 & 3.53 & 3.20 & 3.32 \end{array}$$

Determine the median absolute variability for this data. If there are more than 16 distinct data values, then use the grouped method.

4. A pair of dice, one red and one green, is rolled in the following manner. On the first toss both dice are tossed. Then the red die is tossed while the green die keeps its value to get the second sum. Then the green die is tossed while the red die keeps its value. You continue tossing alternately the red then the green, die and generate the following table.

$$\begin{array}{ccccccc} 8 & 11 & 8 & 6 & 10 & 9 & 9 \\ 7 & 2 & 5 & 10 & 11 & 8 & 5 \\ 3 & 6 & 9 & 7 & 7 & 9 & 10 \\ 7 & 3 & 3 & 7 & 9 & 8 & 7 \\ 5 & 4 & 8 & 11 & 11 & 10 & 10 \end{array}$$

Determine the median absolute variability for this data. If there are more than 16 distinct data values, then use the grouped method.

5. The following random sample of 72 observations comes from a gamma distribution, so $\mu_{min} = 0$ and μ_{max} has no upper limit. The raw data are

2.15	1.92	1.92	2.20	3.67	4.55	3.32	2.71	4.87
2.09	1.13	1.09	2.12	4.89	3.03	3.17	4.74	4.44
4.61	2.78	2.97	2.78	2.88	3.66	2.55	4.40	3.54
2.18	3.43	3.84	4.60	1.76	2.85	2.27	1.38	1.24
4.22	1.67	4.51	3.50	6.69	5.92	6.42	0.95	4.57
3.10	3.26	2.95	2.45	1.95	2.06	3.10	2.03	4.98
0.86	2.61	1.92	2.78	1.01	2.64	3.14	0.86	3.68
2.71	3.06	4.33	4.63	2.47	2.92	1.38	2.81	5.71

Determine the median absolute variability for this data. If there are more than 16 distinct data values, then use the grouped method.

6.4 Average Absolute Deviation

As WE HAVE SEEN, the **variability** of data estimates how far apart two randomly chosen observations will be. A **deviation** uses a fixed point, x, and determines the average distance a randomly chosen point will be to x. Just as with variability, we can talk about average absolute deviation from x, or median absolute deviation from x. In this section we look at the average, denoted **aad**(x). In the next section we will look at the median, denoted **mad**(x).

There are four commonly used points from which the deviations are calculated: zero (the origin), x_{mid} (the midrange), M (the median) — although there is more than one type of median which could be used here — and $\bar{x}$ (the mean). By far, the last two are more common than the first two.

Average Absolute Central Deviation

THE AVERAGE absolute deviation about zero is the **average absolute central deviation**. When all of the data are positive, this is simply another name for the mean, because

$$\left| x_{(i)} - 0 \right| = \left| x_{(i)} \right| = x_{(i)}.$$

Thus, this is more meaningful when the data can be positive or negative.

Example 1:
A number line is stretched across the floor with a scale from -5 to 5. A mechanism is set up to drop a dart onto this line at random intervals in such a way that each point along the number line is equally likely to get hit by the dart. As each dart hits, you record the position along the line to three decimal places. The following table is generated by dropping 40 darts.

3.093	3.814	-3.303	-2.465	-2.357	-3.135	-1.806	-4.250
3.229	4.293	-1.830	1.936	2.739	-0.295	-3.660	-3.716
2.437	3.097	-4.847	3.239	4.093	-0.369	0.513	0.569
-2.507	-0.007	2.689	-1.671	-0.330	-3.288	-2.374	1.734
2.555	0.167	1.963	-1.761	-0.336	-0.763	3.442	3.788

Determine aad(0) for this data.

Answer:
Subtracting 0 does not affect the magnitude, so all we need to do is take the absolute value for each observation. This gives (after sorting)

0.007	0.167	0.295	0.330	0.336	0.369	0.513	0.569
0.763	1.671	1.734	1.761	1.806	1.830	1.936	1.963
2.357	2.374	2.437	2.465	2.507	2.555	2.689	2.739
3.093	3.097	3.135	3.229	3.239	3.288	3.303	3.442
3.660	3.716	3.788	3.814	4.093	4.250	4.293	4.847

6.4. AVERAGE ABSOLUTE DEVIATION

We observe that there are no repeated values. The mean of these absolute values will be the average absolute deviation about 0. Hence,

$$\text{aad}(0) = \frac{0.007 + \cdots + 4.847}{40} = \frac{94.460}{40} \frac{4723}{2000} = 2.3615.$$

Average Absolute Midpoint Deviation

THE AVERAGE absolute deviation about the midrange is the **average absolute midpoint deviation**. For this, we subtract the midrange of the data from every observation and then take the absolute value of each difference. The mean of these absolute differences is the statistic sought.

Example 2:
Determine the midrange, x_{mid}, for the two-successes data and calculate $\text{aad}(x_{\text{mid}})$.

Answer:
The tally table for this data appears below:

2: 9   4: 15   6: 9   8: 4   10: 6   12: 2   15: 1   17: 1
3: 16 5: 17   7: 4   9: 5   11: 4   14: 5   16: 1   28: 1

From this, we calculate

$$x_{\text{mid}} = \frac{2 + 28}{2} = 15.$$

Subtracting 15 from each observation gives

−13: 9   −11: 15   −9: 9   −7: 4   −5: 6   −3: 2   0: 1   2: 1
−12: 16 −10: 17   −8: 4   −6: 5   −4: 4   −1: 5   1: 1   13: 1

Taking absolute values and combining results with the same value we get the following frequency table of absolute differences.

| 0: 1 | 2: 3 | 4: 4 | 6: 5 | 8: 4 | 10: 17 | 12: 16 |
| 1: 6 | 3: 2 | 5: 6 | 7: 4 | 9: 9 | 11: 15 | 13: 10 |

The mean of this is the statistic sought. Hence,

$$\text{aad}(15) = \frac{1 \cdot 6 + \cdots + 13 \cdot 10}{100} = \frac{892}{100} = \frac{223}{25} = 8.92.$$

Average Absolute Deviation About the Median

THE AVERAGE ABSOLUTE deviation is usually calculated about the median. That is what we will calculate here.

Example 3:
Determine the median, M, for the two-success data and calculate $\text{aad}(M)$.

Answer:
The tally table for this data appears below:

| 2: 9 | 4: 15 | 6: 9 | 8: 4 | 10: 6 | 12: 2 | 15: 1 | 17: 1 |
| 3: 16 | 5: 17 | 7: 4 | 9: 5 | 11: 4 | 14: 5 | 16: 1 | 28: 1 |

By the median-of-medians method $\#(M) = (1+100)/2 = 50.5$. By the traditional sample method $\#(M) = 100/2 = 50$. In either case, we get the same result for the median; namely,

$$M = \frac{x_{(50)} + x_{(51)}}{2} = 5$$

as $x_{(50)} = x_{(51)} = 5$. Subtracting 5 from each observation gives the following table of differences.

| -3: 9 | -1: 15 | 1: 9 | 3: 4 | 5: 6 | 7: 2 | 10: 1 | 12: 1 |
| -2: 16 | 0: 17 | 2: 4 | 4: 5 | 6: 4 | 9: 5 | 11: 1 | 23: 1 |

Taking absolute values and combining results with the same value gives

6.4. AVERAGE ABSOLUTE DEVIATION

$$\begin{array}{llllllll}
0:\ 17 & 2:\ 20 & 4:\ 5 & 6:\ 4 & 9:\ 5 & 11:\ 1 & 23:\ 1 \\
1:\ 26 & 3:\ 13 & 5:\ 6 & 7:\ 2 & 10:\ 1 & 12:\ 1 &
\end{array}$$

The mean of the above values is the statistic sought.

$$\text{aad}(5) = \frac{1 \cdot 26 + \cdots + 23 \cdot 1}{100} = \frac{294}{100} = \frac{147}{50} = 2.94.$$

The restricted proper method begins by determining the effective addresses of the sample means.

$$\#(x_{\min}) = \frac{1+9}{2} = 5 \text{ and } \#(x_{\max}) = 100.$$

Whence,

$$\#(M) = \frac{5 + 100}{2} = 52.5.$$

Now, $x_{(52)} = 5$ and $\#(5) = (41+57)/2 = 49 < 52.5$; thus, $y_1 = 5$, $m_1 = 49$, $y_2 = 6$ and $m_2 = (58+66)/2 = 62$. By interpolation,

$$M = \frac{5 \cdot (62 - 52.5) + 6 \cdot (52.5 - 49)}{62 - 49} = \frac{47.5 + 21}{13} = \frac{137}{26} \approx 5.2692.$$

The following difference table shows only numerators as all denominators are 26.

$$\begin{array}{llllllll}
-85:\ 9 & -33:\ 15 & 19:\ 9 & 71:\ 4 & 123:\ 6 & 175:\ 2 & 253:\ 1 & 305:\ 1 \\
-59:\ 16 & -7:\ 17 & 45:\ 4 & 97:\ 5 & 149:\ 4 & 227:\ 5 & 279:\ 1 & 591:\ 1
\end{array}$$

Taking the absolute value and sorting the values from smallest to largest, the numerators are

$$\begin{array}{llllllll}
7:\ 17 & 33:\ 15 & 59:\ 16 & 85:\ 9 & 123:\ 6 & 175:\ 2 & 253:\ 1 & 305:\ 1 \\
19:\ 9 & 45:\ 4 & 71:\ 4 & 97:\ 5 & 149:\ 4 & 227:\ 5 & 279:\ 1 & 591:\ 1
\end{array}$$

The average of the above, remembering to divide by 26, the hidden denominator, will be the average absolute deviation about the median.

$$\text{aad}(137/26) = \frac{7 \cdot 17 + \cdots + 591 \cdot 1}{2600} = \frac{7690}{2600} = \frac{769}{260} \approx 2.9577.$$

Average Absolute Deviation About the Mean

ALMOST AS COMMON as the average absolute deviation about the median is the average absolute difference about the mean. Although we use aad($\bar{x}$) for this, many people use aad(μ).

Example 4:
Determine the mean, $\bar{x}$, for the two-successes data and calculate aad($\bar{x}$).

Answer:
The tally table for this data appears below:

| 2: 9 | 4: 15 | 6: 9 | 8: 4 | 10: 6 | 12: 2 | 15: 1 | 17: 1 |
| 3: 16 | 5: 17 | 7: 4 | 9: 5 | 11: 4 | 14: 5 | 16: 1 | 28: 1 |

The mean for this data is

$$\bar{x} = \frac{2 \cdot 9 + \cdots + 28 \cdot 1}{100} = \frac{644}{100} = \frac{161}{25} = 6.44.$$

| -4.44: 9 | -2.44: 15 | -0.44: 9 | 1.56: 4 | 3.56: 6 | 5.56: 2 | 8.56: 1 | 10.56: 1 |
| -3.44: 16 | -1.44: 17 | 0.56: 4 | 2.56: 5 | 4.56: 4 | 7.56: 5 | 9.56: 1 | 21.56: 1 |

Taking absolute values and sorting from smallest to largest gives

| 0.44: 9 | 1.44: 17 | 2.44: 15 | 3.44: 16 | 4.44: 9 | 5.56: 2 | 8.56: 1 | 10.56: 1 |
| 0.56: 4 | 1.56: 4 | 2.56: 5 | 3.56: 6 | 4.56: 4 | 7.56: 5 | 9.56: 1 | 21.56: 1 |

The mean of the above values is the statistic sought.

$$\text{aad}(6.44) = \frac{0.44 \cdot 9 + \cdots + 21.56 \cdot 1}{100} = \frac{320.08}{100} = \frac{4001}{1250} = 3.2008.$$

Grouped Data

ALL OF THESE statistics can be done from grouped data by using the midpoints of the intervals as the observations and the number

6.4. AVERAGE ABSOLUTE DEVIATION

of observations per interval as the frequency for that midpoint. We will only show an example for the average absolute deviation about the median, but the others are done in a similar manner.

Example 5:
Suppose the random gamma data is partitioned into five intervals and calculate the average absolute deviation about the median for this data.

Answer:
   The width of each interval is

$$w = \frac{31 - 4}{5} = 5.4.$$

This gives the following intervals, midpoints and their frequencies:

I_i	[4, 9.4)	[9.4, 14.8)	[14.8, 20.2)	[20.2, 25.6)	[25.6, 31]
m_i	6.7	12.1	17.5	22.9	28.3
f_i	8	7	4	4	2

There are 25 observations.
   By the median of medians method,

$$\#(M) = \frac{1 + 25}{2} = 13$$

and $M = x_{(13)} = 12.1$. The traditional sample method gives

$$\#(M) = \frac{25}{2} = 12.5$$

but agrees with the median-of-medians method on the value of M. Subtracting this value from each midpoint gives the following tally table:

$$-5.4{:}\ 8 \quad 0{:}\ 7 \quad 5.4{:}\ 4 \quad 10.8{:}\ 4 \quad 16.2{:}\ 2$$

Taking the absolute value of each difference, combining those of the same magnitude and sorting from smallest to largest gives

$$0: 7 \quad 5.4: 12 \quad 10.8: 4 \quad 16.2: 2$$

The mean of the above is the statistic sought; hence,

$$\mathrm{aad}(5.4) = \frac{5.4 \cdot 12 + 10.8 \cdot 4 + 16.2 \cdot 2}{25} = \frac{702}{125} = 5.616.$$

The restricted proper method uses the effective position of the extremes; hence,

$$\#(x_{\min}) = \frac{1+8}{2} = 4.5 \text{ and } \#(x_{\max}) = \frac{24+25}{2} = 24.5.$$

This makes

$$\#(M) = \frac{4.5 + 24.5}{2} = 14.5.$$

Now $x_{(14)} = 12.1$ and $\#(12.1) = (9+15)/2 = 12 < 14.5$; so $y_1 = 12.1$, $m_1 = 12$, $y_2 = 17.5$ and $m_2 = (16+19)/2 = 17.5$. By interpolation,

$$M = \frac{12.1 \cdot (17.5 - 14.5) + 17.5(14.5 - 12)}{17.5 - 12}$$
$$= \frac{36.3 + 43.75}{5.5} = \frac{1601}{110} \approx 14.5545.$$

Subtracting M from each value gives the following table

$$-432/55: 8 \quad -27/11: 7 \quad 162/55: 4 \quad 459/55: 4 \quad 756/55: 2$$

Taking absolute values and sorting them from smallest to largest gives

$$27/11: 7 \quad 162/55: 4 \quad 432/55: 8 \quad 459/55: 4 \quad 756/55: 2$$

The mean of the above is the statistic sought; hence,

$$\mathrm{aad}\left(\frac{1601}{110}\right) = \frac{(27/11) \cdot 7 + \cdots + (756/55) \cdot 2}{25} = \frac{8397}{1375} \approx 6.1069.$$

6.4. AVERAGE ABSOLUTE DEVIATION

Exercises 6.4

1. One day while walking through a field you notice there seem to be a lot of stones lying on the ground. Curious about what the average size of a stone is in the field you randomly collect 30 stones. You weigh each stone and record its weight in grams. The following table shows the raw data.

110	126	75	110	126	94
90	112	111	139	131	92
102	103	96	75	67	101
104	111	100	65	92	93
106	104	93	115	109	95

   Determine the average absolute central deviation, average absolute midpoint deviation, average absolute deviation about the median and average absolute deviation about the mean for this data. Then group the data into five intervals and repeat the calculations.

2. The scores, in the order graded, on a recent midterm examination in Statistics, a class with 30 students, are given below.

70	80	99	98	85	89
87	79	83	38	69	70
60	69	78	40	75	56
70	51	99	69	95	86
57	53	47	50	55	81

   Determine the average absolute central deviation, average absolute midpoint deviation, average absolute deviation about the median and average absolute deviation about the mean for this data. Then group the data into five intervals and repeat the calculations.

3. Below is a sample of the monthly salary for 30 randomly selected people in mid-management in thousands of dollars.

3.08	3.18	3.13	7.41	3.73	3.05
4.17	3.58	3.36	3.27	3.36	4.74
3.32	3.61	4.26	3.02	3.45	4.06
6.73	4.72	3.13	3.15	3.70	3.59
3.12	3.61	5.03	3.53	3.20	3.32

Determine the average absolute central deviation, average absolute midpoint deviation, average absolute deviation about the median and average absolute deviation about the mean for this data. Then group the data into five intervals and repeat the calculations.

4. A pair of dice, one red and one green, is rolled in the following manner. On the first toss both dice are tossed. Then the red die is tossed while the green die keeps its value to get the second sum. Then the green die is tossed while the red die keeps its value. You continue tossing alternately the red then the green, die and generate the following table.

8	11	8	6	10	9	9
7	2	5	10	11	8	5
3	6	9	7	7	9	10
7	3	3	7	9	8	7
5	4	8	11	11	10	10

Determine the average absolute central deviation, average absolute midpoint deviation, average absolute deviation about the median and average absolute deviation about the mean for this data. Then group the data into five intervals and repeat the calculations.

5. The following random sample of 72 observations comes from a gamma distribution, so $\mu_{min} = 0$ and μ_{max} has no upper limit. The raw data are

2.15	1.92	1.92	2.20	3.67	4.55	3.32	2.71	4.87
2.09	1.13	1.09	2.12	4.89	3.03	3.17	4.74	4.44
4.61	2.78	2.97	2.78	2.88	3.66	2.55	4.40	3.54
2.18	3.43	3.84	4.60	1.76	2.85	2.27	1.38	1.24
4.22	1.67	4.51	3.50	6.69	5.92	6.42	0.95	4.57
3.10	3.26	2.95	2.45	1.95	2.06	3.10	2.03	4.98
0.86	2.61	1.92	2.78	1.01	2.64	3.14	0.86	3.68
2.71	3.06	4.33	4.63	2.47	2.92	1.38	2.81	5.71

Determine the average absolute central deviation, average absolute midpoint deviation, average absolute deviation about the median and average absolute deviation about the mean for this data. Then group the data into seven intervals and repeat the calculations.

6.5 Median Absolute Deviation

Being an average the average absolute deviation is affected by the presence of outliers. The **median absolute deviation** is robust so it does not suffer from this problem. As with all deviations, there is a fixed point x about which the differences are measured. In the last section we took the arithmetic average of these differences. In this section we will take the median of the differences about x. The notation for **median absolute deviation about** x is **mad(x)**.

The same four commonly used points for aad(x) are used for mad(x); namely, zero (the origin), x_{mid} (the midrange), M (the median) and $\bar{x}$ () the mean.

Median Absolute Central Deviation

The median absolute deviation about zero is the **median absolute central deviation**. When all of the data are positive, this is simply another name for the median, because

$$\left| x_{(i)} - 0 \right| = \left| x_{(i)} \right| = x_{(i)}.$$

Thus, this is more meaningful when the data can be positive or negative.

Example 1:
A number line is stretched across the floor with a scale from -5 to 5. A mechanism is set up to drop a dart onto this line at random intervals in such a way that each point along the number line is equally likely to get hit by the dart. As each dart hits, you record

the position along the line to three decimal places. The following table is generated by dropping 40 darts.

3.093	3.814	−3.303	−2.465	−2.357	−3.135	−1.806	−4.250
3.229	4.293	−1.830	1.936	2.739	−0.295	−3.660	−3.716
2.437	3.097	−4.847	3.239	4.093	−0.369	0.513	0.569
−2.507	−0.007	2.689	−1.671	−0.330	−3.288	−2.374	1.734
2.555	0.167	1.963	−1.761	−0.336	−0.763	3.442	3.788

Determine mad(0) *for this data.*

Answer:

Subtracting 0 does not affect the magnitude, so all we need to do is take the absolute value for each observation. This gives (after sorting)

0.007	0.167	0.295	0.330	0.336	0.369	0.513	0.569
0.763	1.671	1.734	1.761	1.806	1.830	1.936	1.963
2.357	2.374	2.437	2.465	2.507	2.555	2.689	2.739
3.093	3.097	3.135	3.229	3.239	3.288	3.303	3.442
3.660	3.716	3.788	3.814	4.093	4.250	4.293	4.847

We observe that there are no repeated values in this list, so all three methods will agree on the median. There are 40 data, so the median is

$$\mathrm{mad}(0) = \frac{x_{(20)} + x_{(21)}}{2} = \frac{2.465 + 2.507}{2} = 2.486.$$

Median Absolute Midpoint Deviation

THE MEDIAN absolute deviation about the midrange is the **median absolute midpoint deviation**. For this, we subtract the midrange of the data from every observation and then take the absolute value of each difference. The median of these absolute differences is the statistic sought.

6.5. MEDIAN ABSOLUTE DEVIATION

Example 2:
Determine the midrange, x_{mid}, for the two-successes data and calculate $\text{mad}(x_{\text{mid}})$.

Answer:
The tally table for this data appears below:

2: 9	4: 15	6: 9	8: 4	10: 6	12: 2	15: 1	17: 1
3: 16	5: 17	7: 4	9: 5	11: 4	14: 5	16: 1	28: 1

From this, we calculate
$$x_{\text{mid}} = \frac{2+28}{2} = 15.$$

Subtracting 15 from each observation, taking absolute values, and combining results with the same value gives

0: 1	2: 3	4: 4	6: 5	8: 4	10: 17	12: 16
1: 6	3: 2	5: 6	7: 4	9: 9	11: 15	13: 10

Although the median-of-medians and traditional sample method calculate the effective address differently, both agree that
$$\text{mad}(15) = \frac{d_{(50)} + d_{(51)}}{2} = \frac{10+10}{2} = 10.$$

In the restricted proper method we first observe that
$$\#(d_{\min}) = 1 \text{ and } \#(d_{\max}) = \frac{91+100}{2} = 95.5.$$

We conclude that
$$\#\bigl(\text{mad}(15)\bigr) = \frac{1+95.5}{2} = \frac{193}{4} = 48.25.$$

Now, $d_{(48)} = 10$ and $\#(10) = (45+61)/2 = 53 > 48.25$. Whence, $y_2 = 10$, $m_2 = 53$, $y_1 = 9$ and $m_1 = (36+44)/2 = 40$. By interpolation,
$$\text{mad}(15) = \frac{9 \cdot (53 - 48.25) + 10 \cdot (48.25 - 40)}{53 - 40}$$
$$= \frac{42.75 + 82.5}{13} = \frac{501}{52} \approx 9.6346.$$

Median Absolute Deviation About the Median

THE MEDIAN ABSOLUTE deviation is usually calculated about the median. That is what we will calculate here.

Example 3:
Determine the median, M, for the two-successes data and calculate $\text{mad}(M)$.

Answer:
The tally table for this data appears below:

2: 9	4: 15	6: 9	8: 4	10: 6	12: 2	15: 1	17: 1
3: 16	5: 17	7: 4	9: 5	11: 4	14: 5	16: 1	28: 1

By the median-of-medians method $\#(M) = (1+100)/2 = 50.5$. By the traditional sample method $\#(M) = 100/2 = 50$. In either case, we get the same result for the median; namely,

$$M = \frac{x_{(50)} + x_{(51)}}{2} = 5$$

as $x_{(50)} = x_{(51)} = 5$. Subtracting 5 from each observation, taking absolute values and combining results having the same value, gives the following table of absolute differences.

0: 17	2: 20	4: 5	6: 4	9: 5	11: 1	23: 1
1: 26	3: 13	5: 6	7: 2	10: 1	12: 1	

The median-of-medians method and the traditional sample method will again agree on the median. Therefore,

$$\text{mad}(5) = \frac{d_{(50)} + d_{(51)}}{2} = \frac{2+2}{2} = 2.$$

The restricted proper method begins by determining the effective addresses of the sample means.

$$\#(x_{\min}) = \frac{1+9}{2} = 5 \text{ and } \#(x_{\max}) = 100.$$

6.5. MEDIAN ABSOLUTE DEVIATION

Whence,
$$\#(M) = \frac{5+100}{2} = 52.5.$$

Now, $x_{(52)} = 5$ and $\#(5) = (41+57)/2 = 49 < 52.5$; thus, $y_1 = 5$, $m_1 = 49$, $y_2 = 6$ and $m_2 = (58+66)/2 = 62$. By interpolation,

$$M = \frac{5 \cdot (62 - 52.5) + 6 \cdot (52.5 - 49)}{62 - 49} = \frac{47.5 + 21}{13} = \frac{137}{26} \approx 5.2692.$$

The following difference table shows only numerators as all denominators are 26 — again we have taken absolute values and sorted from smallest to largest.

7: 17	33: 15	59: 16	85: 9	123: 6	175: 2	253: 1	305: 1
19: 9	45: 4	71: 4	97: 5	149: 4	227: 5	279: 1	591: 1

We will remember to divide the result by an additional 26 when we calculate the median of the above. The effective address of the extremes are

$$\#(d_{\min}) = \frac{1+17}{2} = 9 \text{ and } \#(d_{\max}) = 100.$$

From which
$$\#\bigl(\text{mad}(137/26)\bigr) = \frac{9+100}{2} = 54.5.$$

Now, $d_{(54)} = 59/26$ and $\#(54/26) = (46+61)/2 = 53.5 < 54.5$. Thereby, $y_1 = 59/26$, $m_1 = 54.5$, $y_2 = 71/26$ and $m_2 = (62+65)/2 = 63.5$. By interpolation,

$$\text{mad}(137/26) = \frac{59 \cdot (63.5 - 54.5) + 71 \cdot (54.5 - 53.5)}{26 \cdot (63.5 - 53.5)}$$
$$= \frac{531 + 71}{260} = \frac{301}{130} \approx 2.3154.$$

Median Absolute Deviation About the Mean

ALMOST AS COMMON as the median absolute deviation about the median is the median absolute difference about the mean. Although we use $\text{mad}(\bar{x})$ for this, many people use $\text{mad}(\mu)$.

Example 4:
Calculate the mean, $\bar{x}$ and $\text{mad}(\bar{x})$ for the two-success data.

Answer:
The tally table for this data appears below:

2: 9   4: 15   6: 9   8: 4   10: 6   12: 2   15: 1   17: 1
3: 16   5: 17   7: 4   9: 5   11: 4   14: 5   16: 1   28: 1

The mean for this data is

$$\bar{x} = \frac{2 \cdot 9 + \cdots + 28 \cdot 1}{100} = \frac{644}{100} = \frac{161}{25} = 6.44.$$

Subtracting the mean from each observation, taking absolute values and sorting from smallest to largest gives the following tally table for absolute differences.

0.44: 9   1.44: 17   2.44: 15   3.44: 16   4.44: 9   5.56: 2   8.56: 1   10.56: 1
0.56: 4   1.56: 4   2.56: 5   3.56: 6   4.56: 4   7.56: 5   9.56: 1   21.56: 1

There are 100 data in this table.
Both the median-of-medians method and the traditional sample method conclude

$$\text{mad}(6.44) = \frac{d_{(50)} + d_{(51)}}{2} = \frac{2.56 + 2.56}{2} = 2.56.$$

Grouped Data

ALL OF THESE statistics can be done from grouped data by using the midpoints of the intervals as the observations and the number of observations per interval as the frequency for that midpoint. We will only show an example for the median absolute deviation about the median, but the others are done in a similar manner.

Example 5:
Suppose the random gamma data is partitioned into five intervals and calculate the median absolute deviation about the median for this data.

Answer:
  The width of each interval is
$$w = \frac{31 - 4}{5} = 5.4.$$

  This gives the following intervals, midpoints and their frequencies:

I_i	[4, 9.4)	[9.4, 14.8)	[14.8, 20.2)	[20.2, 25.6)	[25.6, 31]
m_i	6.7	12.1	17.5	22.9	28.3
f_i	8	7	4	4	2

There are 25 observations.
  By the median of medians method,
$$\#(M) = \frac{1 + 25}{2} = 13$$
and $M = x_{(13)} = 12.1$. The traditional sample method gives
$$\#(M) = \frac{25}{2} = 12.5$$
but agrees with the median-of-medians method on the value of M. Subtracting this value from each midpoint, taking absolute values and combining results with the same value gives the following tally table:

0: 7   5.4: 12   10.8: 4   16.2: 2

The median of the above has the same address as before; whence,

$$\mathrm{mad}(12.1) = d_{(13)} = 5.4.$$

The restricted proper method uses the effective position of the extremes; hence,

$$\#(x_{\min}) = \frac{1+8}{2} = 4.5 \text{ and } \#(x_{\max}) = \frac{24+25}{2} = 24.5.$$

This makes

$$\#(M) = \frac{4.5 + 24.5}{2} = 14.5.$$

Now $x_{(14)} = 12.1$ and $\#(12.1) = (9+15)/2 = 12 < 14.5$; so $y_1 = 12.1$, $m_1 = 12$, $y_2 = 17.5$ and $m_2 = (16+19)/2 = 17.5$. By interpolation,

$$M = \frac{12.1 \cdot (17.5 - 14.5) + 17.5(14.5 - 12)}{17.5 - 12}$$
$$= \frac{36.3 + 43.75}{5.5} = \frac{1601}{110} \approx 14.5545.$$

Subtracting M from each value taking absolute values and sorting them from smallest to largest gives the following table:

27/11: 7   162/55: 4   432/55: 8   459/55: 4   756/55: 2

The effective addresses of the extremes have become

$$\#(d_{\min}) = \frac{1+7}{2} = 4 \text{ and } \#(d_{\max}) = \frac{24+25}{2} = 24.5.$$

From this, the effective address of the median is

$$\#\bigl(\mathrm{mad}(1601/110)\bigr) = \frac{4 + 24.5}{2} = 14.25.$$

6.5. MEDIAN ABSOLUTE DEVIATION

Now, $d_{(14)} = 432/55$ and $\#(432/55) = (12+19)/2 = 15.5 > 14.25$. Wherefrom, $y_2 = 432/55$, $m_2 = 15.5$, $y_1 = 162/55$ and $m_1 = (8+11)/2 = 9.5$. By interpolation,

$$\text{mad}(1601/110) = \frac{162 \cdot (15.5 - 14.25) + 432 \cdot (14.25 - 9.5)}{55 \cdot (15.5 - 9.5)}$$
$$= \frac{202.5 + 2052}{330} = \frac{1503}{220} \approx 6.8318.$$

Exercises 6.5

1. One day while walking through a field you notice there seem to be a lot of stones lying on the ground. Curious about what the average size of a stone is in the field you randomly collect 30 stones. You weigh each stone and record its weight in grams. The following table shows the raw data.

110	126	75	110	126	94
90	112	111	139	131	92
102	103	96	75	67	101
104	111	100	65	92	93
106	104	93	115	109	95

Determine the median absolute central deviation, median absolute midpoint deviation, median absolute deviation about the median and median absolute deviation about the mean for this data. Then group the data into five intervals and repeat the calculations.

2. The scores, in the order graded, on a recent midterm examination in Statistics, a class with 30 students, are given below.

70	80	99	98	85	89
87	79	83	38	69	70
60	69	78	40	75	56
70	51	99	69	95	86
57	53	47	50	55	81

Determine the median absolute central deviation, median absolute midpoint deviation, median absolute deviation about the median and median absolute deviation about the mean for this data. Then group the data into five intervals and repeat the calculations.

3. Below is a sample of the monthly salary for 30 randomly selected people in mid-management in thousands of dollars.

3.08	3.18	3.13	7.41	3.73	3.05
4.17	3.58	3.36	3.27	3.36	4.74
3.32	3.61	4.26	3.02	3.45	4.06
6.73	4.72	3.13	3.15	3.70	3.59
3.12	3.61	5.03	3.53	3.20	3.32

Determine the median absolute central deviation, median absolute midpoint deviation, median absolute deviation about the median and median absolute deviation about the mean for this data. Then group the data into five intervals and repeat the calculations.

4. A pair of dice, one red and one green, is rolled in the following manner. On the first toss both dice are tossed. Then the red die is tossed while the green die keeps its value to get the second sum. Then the green die is tossed while the red die keeps its value. You continue tossing alternately the red then the green, die and generate the following table.

8	11	8	6	10	9	9
7	2	5	10	11	8	5
3	6	9	7	7	9	10
7	3	3	7	9	8	7
5	4	8	11	11	10	10

Determine the median absolute central deviation, median absolute midpoint deviation, median absolute deviation about the median and median absolute deviation about the mean for this data. Then group the data into five intervals and repeat the calculations.

5. The following random sample of 72 observations comes from a gamma distribution, so $\mu_{\min} = 0$ and $\mu_{\max}$ has no upper limit. The raw data are

2.15	1.92	1.92	2.20	3.67	4.55	3.32	2.71	4.87
2.09	1.13	1.09	2.12	4.89	3.03	3.17	4.74	4.44
4.61	2.78	2.97	2.78	2.88	3.66	2.55	4.40	3.54
2.18	3.43	3.84	4.60	1.76	2.85	2.27	1.38	1.24
4.22	1.67	4.51	3.50	6.69	5.92	6.42	0.95	4.57
3.10	3.26	2.95	2.45	1.95	2.06	3.10	2.03	4.98
0.86	2.61	1.92	2.78	1.01	2.64	3.14	0.86	3.68
2.71	3.06	4.33	4.63	2.47	2.92	1.38	2.81	5.71

Determine the median absolute central deviation, median absolute midpoint deviation, median absolute deviation about the median and median absolute deviation about the mean for this data. Then group the data into seven intervals and repeat the calculations.

6.6 Root Mean Square Average

THERE IS ONE additional average which is used for deviations. Actually there is a class of averages, but we will only look at the second power.

Let $x_1, \ldots, x_n$ be a collection of values. The **root mean square average**, denoted $\| x \|$, for these values is defined as

$$\| x \| = \sqrt{\frac{\sum_{i=1}^{n} x_i^2}{n}}.$$

Thus, we square each value, average them, and then take the square root to get the result. When there is repetition the formula looks slightly different. Suppose there are k distinct values $y_1, \ldots, y_k$ and that $\text{frq}(y_i) = f_i$ $(i = 1, \ldots, k)$. Then

$$\|y\| = \sqrt{\frac{\sum_{i=1}^{k} y_i^2 \cdot f_i}{\sum_{i=1}^{k} f_i}}.$$

Example 1:
A number line is stretched across the floor with a scale from -5 to 5. A mechanism is set up to drop a dart onto this line at random intervals in such a way that each point along the number line is equally likely to get hit by the dart. As each dart hits, you record the position along the line to three decimal places. The following table is generated by dropping 40 darts.

3.093	3.814	−3.303	−2.465	−2.357	−3.135	−1.806	−4.250
3.229	4.293	−1.830	1.936	2.739	−0.295	−3.660	−3.716
2.437	3.097	−4.847	3.239	4.093	−0.369	0.513	0.569
−2.507	−0.007	2.689	−1.671	−0.330	−3.288	−2.374	1.734
2.555	0.167	1.963	−1.761	−0.336	−0.763	3.442	3.788

Determine the root mean square average for this data.

Answer:
We begin by squaring each observation.

9.566649	14.546596	10.909809	6.076225	5.555449	9.828225	3.261636	18.062500
10.426441	18.429849	3.348900	3.748096	7.502121	0.087025	13.395600	13.808656
5.938969	9.591409	23.493409	10.491121	16.752649	0.136161	0.263169	0.323761
6.285049	0.000049	7.230721	2.792241	0.108900	10.810944	5.635876	3.006756
6.528025	0.027889	3.853369	3.101121	0.112896	0.582169	11.847364	14.348944

6.6. ROOT MEAN SQUARE AVERAGE

The mean of the above squares is

$$\frac{9.566649 + \cdots + 14.348944}{40} = \frac{291.816738}{40} = 7.29541845.$$

Taking the square root of the above we get

$$\|x\| = \sqrt{7.29541845} \approx 2.701003.$$

Theorem 6.1. *Let $n > 1$ and $x_1, \ldots, x_n$ be nonnegative real numbers that are not all zero then*

$$\bar{x} < \|x\|.$$

Proof. Omitted. $\square$

If you compare Example 1, §6.4, page 306 (the absolute value makes all numbers nonnegative and $|x|^2 = x^2$) you will see the theorem holds for these values.

Example 2:
Compare the arithmetic average to the root mean square average of the following values.

$$2 \quad 4 \quad 2 \quad 3 \quad 2 \quad 3 \quad 5 \quad 2 \quad 4 \quad 3$$

Sorting the values from smallest to largest gives the following frequency table.

$$2\colon 4 \quad 3\colon 3 \quad 4\colon 2 \quad 5\colon 1$$

The arithmetic average is

$$\bar{x} = \frac{2 \cdot 4 + \cdots + 5 \cdot 1}{4 + \cdots + 1} = \frac{30}{10} = 3.$$

Squaring each value and calculating the mean of the squares gives

$$\|x\|^2 = \frac{4 \cdot 4 + \cdots + 25 \cdot 1}{10} = \frac{100}{10} = 10.$$

From this,

$$\|x\| = \sqrt{10} \approx 3.16227766.$$

Observe: $\bar{x} < \|x\|$ as claimed.

Exercises 6.6

1. Determine the root mean square average of the following quantities:

   16   93   42   42   51   25   23   93   91   57

2. Determine the root mean square average of the following quantities:

   0.96   1.08   1.03   1.05   1.08   0.96   1.03   1.08   1.03   0.96

3. Determine the root mean square average of the following quantities:

   −3.41   −4.43   −1.62   −2.95   1.62   1.62   −2.95   −4.43   4.78   4.43   3.41   4.78

6.7 Variation and Standard Deviance

THE MOST FORGOTTEN pair of statistics of dispersion is quite similar to the most remembered pair. The pair we are going to talk about is rarely mentioned in books of statistics. In fact, it may be the very similarity it shares with the most well-known pair which causes it to

6.7. VARIATION AND STANDARD DEVIANCE

be the least known. Both pairs are based upon the root mean square average.

The **variation**, denoted $\mathcal{V}$, is the mean square of the deviations from the median. Thus, if $x_{(1)}, x_{(2)}, \ldots, x_{(n)}$ are the sorted data, and M is the median then

$$\mathcal{V} = \frac{\sum_{i=1}^{n}(x_{(i)} - M)^2}{n-1}$$

for the median-of-medians and traditional sample methods and

$$\mathcal{V} = \frac{\sum_{i=1}^{n}(x_{(i)} - M)^2}{n}$$

for the restricted proper method.

The **standard deviance** is the square root of the variation and is denoted s_d. Hence,

$$s_\text{d}^2 = \mathcal{V} \text{ and } s_\text{d} = \sqrt{\mathcal{V}}.$$

The reason for the $n-1$ for the median-of-medians and traditional sample methods is easy to explain. If $n = 2m - 1$ then only $x_{(m)}$ is known, all of the others can be freely chosen. If $n = 2m$ then you can choose $x_{(m)}$ or $x_{(m+1)}$ and immediately know the other, but there is no restriction for the other values. This is because the position of the median depends only upon the number of items and nothing else.

The median of the restricted proper method is much harder to analyze. First the position depends upon the frequencies of the extremes and is not fixed. Second the value depends upon both the values and frequencies of the two distinct values closest to the medial position. This combination of dependencies means that the sample median acts more like a population parameter and less like a sample statistic. Thus, the sample must be treated as if it were a census and we must use n for the denominator.

Example 1:
Determine the variation and standard deviance for the oyster data.

Answer:
   The sorted data are repeated here. There are 50 observations.

```
 5.76   6.10   6.36   6.37   6.42   6.53   6.63   6.67   6.74   6.91
 6.91   6.97   7.03   7.04   7.37   7.40   7.71   7.97   8.02   8.04
 8.12   8.14   8.42   8.53   8.61   8.62   9.01   9.20   9.21   9.39
 9.47   9.53   9.55   9.98  10.24  10.50  10.62  10.63  10.63  10.81
10.81  10.86  11.02  11.09  11.17  11.31  11.59  11.97  12.07  13.18
```

In the median-of-medians and traditional sample methods,

$$M = \frac{x_{(25)} + x_{(26)}}{2} = \frac{8.61 + 8.62}{2} = 8.615.$$

Subtracting the median from each datum and taking absolute values gives

```
2.855 2.515 2.255 2.245 2.195 2.085 1.985 1.945 1.875 1.705
1.705 1.645 1.585 1.575 1.245 1.215 0.905 0.645 0.595 0.575
0.495 0.475 0.195 0.085 0.005 0.005 0.395 0.585 0.595 0.775
0.855 0.915 0.935 1.365 1.625 1.885 2.005 2.015 2.015 2.195
2.195 2.245 2.405 2.475 2.555 2.695 2.975 3.355 3.455 4.565
```

Squaring and sorting gives

6.7. VARIATION AND STANDARD DEVIANCE

0.000025	0.000025	0.007225	0.038025	0.156025
0.225625	0.245025	0.330625	0.342225	0.354025
0.354025	0.416025	0.600625	0.731025	0.819025
0.837225	0.874225	1.476225	1.550025	1.863225
2.480625	2.512225	2.640625	2.706025	2.907025
2.907025	3.515625	3.553225	3.783025	3.940225
4.020025	4.060225	4.060225	4.347225	4.818025
4.818025	4.818025	5.040025	5.040025	5.085025
5.784025	6.125625	6.325225	6.528025	7.263025
8.151025	8.850625	11.256025	11.937025	20.839225

There are only a few repeated values. Dividing the sum of these by $50 - 1 = 49$ gives

$$\mathcal{V} = \frac{0.000025 \cdot 2 + \cdots + 20.839225}{49}$$
$$= \frac{181.33285}{49} = \frac{3626657}{980000} \approx 3.7006704.$$

Whence, taking the square root,

$$s_{\mathrm{d}} \approx 1.923713.$$

Both $x_{\min}$ and $x_{\max}$ are unique. In addition, both $x_{(25)}$ and $x_{(26)}$ are unique. Therefore, the restricted proper method agrees with the median-of-medians and traditional proper method on the value of M. The only difference is that we divide by 50 rather than 49; so,

$$\mathcal{V} = \frac{0.000025 \cdot 2 + \cdots + 20.839225}{49}$$
$$= \frac{181.33285}{50} = \frac{3626657}{1000000} = 3.626657.$$

From which we get

$$s_{\mathrm{d}} \approx 1.904378.$$

Suppose there are k distinct values $y_1, y_2, \ldots, y_k$ and that for $i = 1, \ldots, k$ $\mathrm{frq}(y_i) = f_i$. Furthermore, suppose M is the median.

Then, with
$$n = \sum_{i=1}^{k} f_i$$
we get
$$\mathcal{V} = \frac{\sum_{i=1}^{n}(y_i - M)^2 \cdot f_i}{n-1}$$
for the median-of-medians and traditional sample methods and
$$\mathcal{V} = \frac{\sum_{i=1}^{n}(y_i - M)^2 \cdot f_i}{n}$$
for the restricted proper method.

Example 2:
Determine the variation and standard deviance for the two-success data.

Answer:
In Example 3, §6.4, starting on page 308 we determined the median for the median-of-medians and traditional method to be $M = 5$. In addition, the absolute deviations from the median were calculated and are repeated below.

| 0: 17 | 2: 20 | 4: 5 | 6: 4 | 9: 5 | 11: 1 | 23: 1 |
| 1: 24 | 3: 13 | 5: 6 | 7: 2 | 10: 1 | 12: 1 | |

Squaring each of the above and dividing by $100 - 1 = 99$ gives
$$\mathcal{V} = \frac{1^2 \cdot 26 + \cdots + 23^2 \cdot 1}{99} = \frac{1992}{99} = \frac{664}{33} \approx 20.12121212.$$

6.7. VARIATION AND STANDARD DEVIANCE

Hence,
$$s_d = \sqrt{\frac{1992}{99}} = \frac{2\sqrt{5478}}{33} \approx 4.485667.$$

For the restricted proper method the median is $M = 137/26$. The absolute deviations from the median for this are listed below.

7: 17	33: 15	59: 16	85: 9	123: 6	175: 2	253: 1	305: 1
19: 9	45: 4	71: 4	97: 5	149: 4	227: 5	279: 1	591: 1

where we have again not written the common denominator of 26. Squaring each of the above and dividing by $26^2 \cdot 100$ makes

$$V = \frac{7^2 \cdot 17 + \cdots + 591^2 \cdot 1}{26^2 \cdot 100} = \frac{1299076}{67600} = \frac{324769}{16900} \approx 19.21710059.$$

Therefore,
$$s_d \approx 4.383731.$$

Grouped Data

THE TECHNIQUES used for grouped data are the same as we have done before — the midpoints of the intervals are used as observations and the number of observations per interval are used as the frequencies for these midpoints.

Example 3:
A number line is stretched across the floor with a scale from -5 to 5. A mechanism is set up to drop a dart onto this line at random intervals in such a way that each point along the number line is equally likely to get hit by the dart. As each dart hits, you record the position along the line to three decimal places. The following table is generated by dropping 40 darts.

3.093	3.814	−3.303	−2.465	−2.357	−3.135	−1.806	−4.250
3.229	4.293	−1.830	1.936	2.739	−0.295	−3.660	−3.716
2.437	3.097	−4.847	3.239	4.093	−0.369	0.513	0.569
−2.507	−0.007	2.689	−1.671	−0.330	−3.288	−2.374	1.734
2.555	0.167	1.963	−1.761	−0.336	−0.763	3.442	3.788

Group the data into five intervals and use these to determine the variation and standard deviance.

Answer:

Sorting the above from smallest to largest gives

−4.847	−4.250	−3.716	−3.660	−3.303	−3.288	−3.135	−2.507
−2.465	−2.374	−2.357	−1.830	−1.806	−1.761	−1.671	−0.763
−0.369	−0.336	−0.330	−0.295	−0.007	0.167	0.513	0.569
1.734	1.936	1.963	2.437	2.555	2.689	2.739	3.093
3.097	3.229	3.239	3.442	3.788	3.814	4.093	4.293

From the above, we see

$$x_{\min} = {}^-4.847 \text{ and } x_{\max} = 4.293;$$

so, $R = 4.293 + 4.847 = 9.140$.

With five intervals the common width for each interval is

$$w = \frac{9.140}{5} = 1.828.$$

We will round this up to the nearest 0.02 to make $w = 1.84$ and adjust the starting point to be nice. The original value for t_0 is

$$t_0 = \frac{{}^-4.847 + 4.293 - 5 \cdot 1.84}{2} = {}^-4.877$$

and will make the new $t_0 = {}^-4.88$ but keep the same interval width. Observe: $^-4.88 + 5 \cdot 1.84 = 4.32 > x_{\max}$ so everything is covered as it needs to be.

The resulting intervals, their midpoints and the number of observations in each is shown below.

6.7. VARIATION AND STANDARD DEVIANCE

I_i	[-4.88,-3.04)	[-3.04,-1.2)	[-1.2, 0.64)	[0.64, 2.48)	[2.48, 4.32]
m_i	-3.96	-2.12	-0.28	1.56	3.40
f_i	7	8	9	4	12

By the median-of-medians and traditional sample methods the median of the above is

$$M = \frac{m_{(20)} + m_{(21)}}{2} = \frac{-0.28 - 0.28}{2} = -0.28.$$

Subtracting the median from each midpoint, taking the absolute value,[1] and sorting from smallest to largest gives the following absolute difference table.

$$0: 9 \quad 1.84: 12 \quad 3.68: 19$$

From the above we get

$$V = \frac{1.84^2 \cdot 12 + 3.68^2 \cdot 19}{40 - 1} = \frac{297.9328}{39} = \frac{186208}{24375} \approx 7.63930256,$$

which makes

$$s_{\mathrm{d}} = 2.763929.$$

The restricted proper method meeds the effective addresses for the sample means.

$$\#(m_{\min}) = \frac{1+7}{2} = 4 \text{ and } \#(m_{\max}) = \frac{29+40}{2} = 34.5.$$

This makes

$$\#(M) = \frac{4 + 34.5}{2} = \frac{77}{4} = 19.25.$$

Now, $m_{(19)} = {}^-0.28$ and $\#({}^-0.28) = \frac{16+24}{2} = 20 > 19.25$. We conclude $y_2 = {}^-0.28$, $m_2 = 20$, $y_1 = {}^-2.12$ and $m_1 = (8 + 15)/2 = 11.5$. By

---

[1]Taking the absolute value helps in combining like values for when they are squared.

interpolation,

$$M = \frac{^-212 \cdot (20 - 19.25) + {^-28} \cdot (19.25 - 11.5)}{100 \cdot (20 - 11.5)}$$
$$= -\frac{159 + 217}{850} = -\frac{188}{425} \approx -0.442353.$$

Subtracting the median from each midpoint, taking the absolute value, and sorting from smallest to largest gives the following absolute difference table. [I have made all denominators the same]

69/425: 9   713/425: 8   851/42: 4   1495/425: 7   1633/425: 12

From the above we get

$$\mathcal{V} = \frac{69^2 \cdot 9 + \cdots + 1633^2 \cdot 12}{425^2 \cdot 40}$$
$$= \frac{54652048}{7225000} = \frac{6831506}{903125} \approx 7.56429730;$$

from which,

$$s_d \approx 2.750327.$$

Relative Variability

WHAT IS LARGE? For that matter, what is small? Is a standard deviance of 1000 large? Is a standard deviance of 0.01 small? What if I told you that the median for the standard deviance of 1000 is 1000000? Now does it seem large? What if I told you that the median for the standard deviance of 0.01 is 0.0001? Now, does it seem small. The point is that standard deviance, as an absolute measure, is not really that useful in determining how variable data is about the median. The standard deviance and the median have the same units of measure; thus, comparing them will result in a

6.7. VARIATION AND STANDARD DEVIANCE

unit-less measurement, which can be more useful when you need to compare different samples.

The **relative variability about the median**, denoted $\mathcal{V}_M$, is the absolute value of the ratio of the standard deviance to the median. It exists provided that the median is not zero. Thus, in the first case $\mathcal{V}_M = 1000/1000000 = 0.001$; whereas, in the second case $\mathcal{V}_M = 0.01/0.0001 = 100$. Thus, large is small and small is large because the medians make it so.

Example 4:
Determine the relative variability about the median for the oyster data.

Answer:
In Example 1 of this section we determined the median by the traditional sample method is $M = 8.615$ and standard deviance is $s_\mathrm{d} = 1.923713$. We conclude
$$\mathcal{V}_M \approx \frac{1.923713}{8.615} \approx 0.2233.$$

For the restricted proper method, the median is the same, but the standard deviance is now $s_\mathrm{d} \approx 1.904378$. This makes
$$\mathcal{V}_M \approx \frac{1.904378}{8.615} \approx 0.2211.$$

Example 5:
Determine the relative variability about the median for the two-success data.

Answer:
In Example 2 of this section we determined the median by the traditional sample method is $M = 5$ and standard deviance is $s_\mathrm{d} = 4.487919$.

We conclude
$$\mathcal{V}_M \approx \frac{4.487919}{5} \approx 0.8976.$$
For the restricted proper method, the median is $M = 137/26$ and the standard deviance is now $s_\mathrm{d} \approx 4.383731$. This makes
$$\mathcal{V}_M \approx \frac{4.383731}{5.269231} \approx 0.8319.$$

Exercises 6.7

1. One day while walking through a field you notice there seem to be a lot of stones lying on the ground. Curious about what the average size of a stone is in the field you randomly collect 30 stones. You weigh each stone and record its weight in grams. The following table shows the raw data.

110	126	75	110	126	94
90	112	111	139	131	92
102	103	96	75	67	101
104	111	100	65	92	93
106	104	93	115	109	95

    Determine the variation, standard deviance and relative variability about the median. Then group the data into five intervals and repeat the calculations.

2. The scores, in the order graded, on a recent midterm examination in Statistics, a class with 30 students, are given below.

70	80	99	98	85	89
87	79	83	38	69	70
60	69	78	40	75	56
70	51	99	69	95	86
57	53	47	50	55	81

6.7. VARIATION AND STANDARD DEVIANCE

Determine the variation, standard deviance and relative variability about the median. Then group the data into five intervals and repeat the calculations.

3. Below is a sample of the monthly salary for 30 randomly selected people in mid-management in thousands of dollars.

3.08	3.18	3.13	7.41	3.73	3.05
4.17	3.58	3.36	3.27	3.36	4.74
3.32	3.61	4.26	3.02	3.45	4.06
6.73	4.72	3.13	3.15	3.70	3.59
3.12	3.61	5.03	3.53	3.20	3.32

Determine the variation, standard deviance and relative variability about the median. Then group the data into five intervals and repeat the calculations.

4. A pair of dice, one red and one green, is rolled in the following manner. On the first toss both dice are tossed. Then the red die is tossed while the green die keeps its value to get the second sum. Then the green die is tossed while the red die keeps its value. You continue tossing alternately the red then the green, die and generate the following table.

8	11	8	6	10	9	9
7	2	5	10	11	8	5
3	6	9	7	7	9	10
7	3	3	7	9	8	7
5	4	8	11	11	10	10

Determine the variation, standard deviance and relative variability about the median. Then group the data into five intervals and repeat the calculations.

5. The following random sample of 72 observations comes from a gamma distribution, so $\mu_{min} = 0$ and μ_{max} has no upper limit. The raw data are

2.15	1.92	1.92	2.20	3.67	4.55	3.32	2.71	4.87
2.09	1.13	1.09	2.12	4.89	3.03	3.17	4.74	4.44
4.61	2.78	2.97	2.78	2.88	3.66	2.55	4.40	3.54
2.18	3.43	3.84	4.60	1.76	2.85	2.27	1.38	1.24
4.22	1.67	4.51	3.50	6.69	5.92	6.42	0.95	4.57
3.10	3.26	2.95	2.45	1.95	2.06	3.10	2.03	4.98
0.86	2.61	1.92	2.78	1.01	2.64	3.14	0.86	3.68
2.71	3.06	4.33	4.63	2.47	2.92	1.38	2.81	5.71

Determine the variation, standard deviance and relative variability about the median. Then group the data into seven intervals and repeat the calculations.

6.8 Variance and Standard Deviation

T<small>HIS SECTION</small>, at first glance, will appear to be a repeat of the last. It isn't. The most common pair of dispersion statistics does resembled the least common when using the definition. However, unlike the other pair, this pair has a calculating formula which completely destroys their similarity. We will see that this formula generalizes to other types of statistics.

The **variance**, denoted **V** or **s^2**, is the mean square of the deviations from the mean. Thus, if $x_1, x_2, \ldots, x_n$ are the observations and $\overline{x}$ their mean then

$$V = \frac{\sum_{i=1}^{n}(x_i - \overline{x})^2}{n - 1}.$$

The reason for the $n-1$ is that if you know all but one observation, and you know the mean of all of them, then you must know all of

6.8. VARIANCE AND STANDARD DEVIATION

them because

$$\bar{x} = \frac{1}{n}\sum_{i=1}^{n} x_i; \text{ that is, } n \cdot \bar{x} = \sum_{i=1}^{n} x_i.$$

The **standard deviation** is the square root of the variance and is denoted **s**. Whereby,

$$s^2 = V \text{ and } s = \sqrt{V}$$

When the sample is a census the variance and standard have a slightly different formula. There is also a different notation so you are not unlikely to get confused. The variance is denoted σ^2 — which implies the standard deviation is σ. The formula is

$$\sigma^2 = \frac{\sum_{i=1}^{n}(x_i - \bar{x})^2}{n}.$$

Example 1:
Determine the variance and the standard deviation for the random gamma data.

Answer:
The sorted data are

4	5	5	5	5
8	9	9	10	12
13	13	13	14	14
15	17	17	18	22
22	24	24	27	31

The mean for the above data is

$$\bar{x} = \frac{4 + 5 \cdot 4 + \cdots + 31}{25} = \frac{356}{25} = 14.24.$$

Subtracting this from each observation gives the following frequency table

$$\begin{array}{lllll}
-10.24\text{: }1 & -5.24\text{: }2 & -1.24\text{: }3 & 2.76\text{: }2 & 9.76\text{: }2 \\
-9.24\text{: }4 & -4.24\text{: }1 & -0.24\text{: }2 & 3.76\text{: }1 & 12.76\text{: }1 \\
-6.24\text{: }1 & -2.24\text{: }1 & 0.76\text{: }1 & 7.76\text{: }2 & 16.76\text{: }1
\end{array}$$

Squaring and sorting gives

$$\begin{array}{lllll}
0.0576\text{: }2 & 5.0176\text{: }1 & 17.9776\text{: }1 & 60.2176\text{: }2 & 104.8576\text{: }1 \\
0.5776\text{: }1 & 7.6176\text{: }2 & 27.4576\text{: }2 & 85.3776\text{: }4 & 162.8176\text{: }1 \\
1.5376\text{: }3 & 14.1376\text{: }1 & 38.9376\text{: }1 & 95.2576\text{: }2 & 280.8976\text{: }1
\end{array}$$

The mean of the above is the variance (except the denominator is one less)

$$V = \frac{0.0576 \cdot 2 + \cdots + 280.8976}{25 - 1}$$
$$= \frac{1352.5600}{24} = \frac{16907}{300} \approx 56.35666667,$$

so that

$$s \approx 7.507108.$$

There is a calculating formula for the variance which does not require one to determine $\bar{x}$ beforehand. We will derive it here.

$$\sum_{i=1}^{n}(x_i - \bar{x})^2 = \sum_{i=1}^{n} x_i^2 - 2 \cdot \bar{x} \cdot x_i + \bar{x}^2$$
$$= \sum_{i=1}^{n} x_i^2 - 2 \cdot \bar{x} \cdot \sum_{i=1}^{n} x_i + \sum_{i=1}^{n} \bar{x}^2$$

6.8. VARIANCE AND STANDARD DEVIATION

$$= \sum_{i=1}^{n} x_i^2 - \frac{2}{n}\left(\sum_{i=1}^{n} x_i\right)^2 + n \cdot \left(\frac{1}{n}\sum_{i=1}^{n} x_i\right)^2$$

$$= \sum_{i=1}^{n} x_i^2 - \frac{1}{n}\left(\sum_{i=1}^{n} x_i\right)^2$$

Placing the above into the formula and getting a common denominator for this expression gives the calculating formula

$$V = \frac{n\sum_{i=1}^{n} x_i^2 - \left(\sum_{i=1}^{n} x_i\right)^2}{n \cdot (n-1)}.$$

Example 2:
Determine the variance and standard deviation for the call center data.

Answer:
The sorted data is

24	24	24	24	26	26	27	27	29	29
29	30	30	31	32	33	35	35	36	37
37	38	38	39	40	42	42	43	43	44
45	45	46	47	48	48	49	50	59	66
66	71	72	75	76	79	84	88	91	112

We calculate

$$\sum_{i=1}^{50} x_i = 24 \cdot 4 + \cdots + 112 = 2311$$

and

$$\sum_{i=1}^{50} x_i^2 = 24^2 \cdot 4 + \cdots + 112^2 = 127735.$$

Therefore,
$$V = \frac{50 \cdot 127735 - 2311^2}{50 \cdot 49} = \frac{1046029}{2450} \approx 426.95061224,$$
from which
$$s \approx 20.662783.$$

Suppose there are k distinct values for observations, $y_1, \ldots, y_k$ and that $\text{frq}(y_i) = f_i$ for $i = 1, \ldots, k$. Then

$$V = \frac{n \sum_{i=1}^{k} x_i^2 \cdot f_i - \left(\sum_{i=1}^{k} x_i \cdot f_i\right)^2}{n \cdot (n-1)},$$

where
$$n = \sum_{i=1}^{k} f_i.$$

For a census the formula is

$$\sigma^2 = \frac{n \sum_{i=1}^{k} x_i^2 \cdot f_i - \left(\sum_{i=1}^{k} x_i \cdot f_i\right)^2}{n^2}.$$

Example 3:
Determine the variance and standard deviation for the dice roll data.
Answer:
The frequency table for this data is

2: 1   5: 6   8: 7   11: 1
3: 5   6: 7   9: 5   12: 1
4: 4   7: 8   10: 5

6.8. VARIANCE AND STANDARD DEVIATION

From this, we calculate

$$\sum_{i=2}^{12} i \cdot \text{frq}(i) = 2 \cdot 1 + 3 \cdot 5 + \cdots + 12 \cdot 1 = 335$$

and

$$\sum_{i=2}^{12} i^2 \cdot \text{frq}(i) = 2^2 \cdot 1 + 3^2 \cdot 5 + \cdots + 12^2 \cdot 1 = 2525.$$

Therefore,

$$V = \frac{50 \cdot 2525 - 335^2}{50 \cdot 49} = \frac{14025}{2450} = \frac{561}{98} \approx 5.72448980.$$

Whereby,

$$s \approx 2.392591.$$

Grouped Data

AS WE HAVE SEEN with other statistics, working with grouped data means that the midpoints of the intervals are treated as observations and the counts for each interval are treated as the frequency of the corresponding midpoint. Many people treat the grouped data as if it were a census; thus, not subtracting one in the denominator. However, as the accuracy is less than with the original data, I do not feel that this is the best of ideas and treat it just lie any other sample. The calculation formula allows a compact way of calculating the variance and standard deviation by simply adding two columns and a row to the frequency table — assuming you are organizing the table into columns.

Example 4:
A number line is stretched across the floor with a scale from ⁻5 to 5. A mechanism is set up to drop a dart onto this line at random intervals in such a way that each point along the number line is equally likely to get hit by the dart. As each dart hits, you record the position along the line to three decimal places. The following table is generated by dropping 40 darts.

3.093	3.814	−3.303	−2.465	−2.357	−3.135	−1.806	−4.250
3.229	4.293	−1.830	1.936	2.739	−0.295	−3.660	−3.716
2.437	3.097	−4.847	3.239	4.093	−0.369	0.513	0.569
−2.507	−0.007	2.689	−1.671	−0.330	−3.288	−2.374	1.734
2.555	0.167	1.963	−1.761	−0.336	−0.763	3.442	3.788

Group the data into five intervals and use these to determine the variance and standard deviation.

Answer:
From by Example 3, §6.7, starting on page 333, the common width for each interval was determined to be $w = 1.84$ and the starting point was modified to $t_0 = {-4.88}$. This gave the following intervals, midpoints and counts.

I_i	m_i	f_i	$m_i \cdot f_i$	$m_i^2 \cdot f_i$
$[-4.88, -3.04)$	−3.96	7	−27.72	109.7712
$[-3.04, -1.20)$	−2.12	8	−16.96	35.9552
$[-1.20, 0.64)$	−0.28	9	−2.52	0.7056
$[0.64, 2.48)$	1.56	4	6.24	9.7344
$[2.48, 4.32]$	3.40	12	40.80	138.7200
Totals:		40	−0.16	294.8864

where we have added two additional columns and a totals line. Using the totals line from above,

$$V = \frac{40 \cdot 294.8864 - (-0.16)^2}{40 \cdot 39} = \frac{294.8608}{1560} = \frac{1772}{9375} \approx 0.18901333.$$

Wherefrom

$$s \approx 0.434757.$$

Relative Variability

JUST AS THE STANDARD deviance is not an absolute measure of variability, neither is the standard deviation. The **relative variability about the mean**, denoted V_m, is the absolute value of the ratio of the standard deviation to the mean. It exists provided that the mean is not zero.

Example 5:
Determine the relative variability about the mean for the random gamma data.

Answer:
In Example 1 of this section we determined the mean is $\bar{x} = 14.24$ and the standard deviation is $s \approx 7.507108$. Thereby,

$$V_m \approx \frac{7.507108}{14.24} \approx 0.5272.$$

Example 6:
Determine the relative variability about the median for the call center data.

Answer:
In Example 2 of this section we determined the sum of the data is 2311, from which we determine

$$\bar{x} = \frac{2311}{50} = 46.22.$$

In addition, the standard deviation is $s \approx 20.662783$. We conclude

$$V_m \approx \frac{20.662783}{46.22} \approx 0.4471.$$

Exercises 6.8

1. One day while walking through a field you notice there seem to be a lot of stones lying on the ground. Curious about what the average size of a stone is in the field you randomly collect 30 stones. You weigh each stone and record its weight in grams. The following table shows the raw data.

110	126	75	110	126	94
90	112	111	139	131	92
102	103	96	75	67	101
104	111	100	65	92	93
106	104	93	115	109	95

   Determine the variance, standard deviation and relative variability about the mean. Then group the data into five intervals and repeat the calculations. Calculate the

2. The scores, in the order graded, on a recent midterm examination in Statistics, a class with 30 students, are given below.

70	80	99	98	85	89
87	79	83	38	69	70
60	69	78	40	75	56
70	51	99	69	95	86
57	53	47	50	55	81

   Determine the variance, standard deviation and relative variability about the mean. Then group the data into five intervals and repeat the calculations.

3. Below is a sample of the monthly salary for 30 randomly selected people in mid-management in thousands of dollars.

3.08	3.18	3.13	7.41	3.73	3.05
4.17	3.58	3.36	3.27	3.36	4.74
3.32	3.61	4.26	3.02	3.45	4.06
6.73	4.72	3.13	3.15	3.70	3.59
3.12	3.61	5.03	3.53	3.20	3.32

6.8. VARIANCE AND STANDARD DEVIATION

Determine the variance, standard deviation and relative variability about the mean. Then group the data into five intervals and repeat the calculations.

4. A pair of dice, one red and one green, is rolled in the following manner. On the first toss both dice are tossed. Then the red die is tossed while the green die keeps its value to get the second sum. Then the green die is tossed while the red die keeps its value. You continue tossing alternately the red then the green, die and generate the following table.

8	11	8	6	10	9	9
7	2	5	10	11	8	5
3	6	9	7	7	9	10
7	3	3	7	9	8	7
5	4	8	11	11	10	10

Determine the variance, standard deviation and relative variability about the mean. Then group the data into five intervals and repeat the calculations.

5. The following random sample of 72 observations comes from a gamma distribution, so $\mu_{min} = 0$ and μ_{max} has no upper limit. The raw data are

2.15	1.92	1.92	2.20	3.67	4.55	3.32	2.71	4.87
2.09	1.13	1.09	2.12	4.89	3.03	3.17	4.74	4.44
4.61	2.78	2.97	2.78	2.88	3.66	2.55	4.40	3.54
2.18	3.43	3.84	4.60	1.76	2.85	2.27	1.38	1.24
4.22	1.67	4.51	3.50	6.69	5.92	6.42	0.95	4.57
3.10	3.26	2.95	2.45	1.95	2.06	3.10	2.03	4.98
0.86	2.61	1.92	2.78	1.01	2.64	3.14	0.86	3.68
2.71	3.06	4.33	4.63	2.47	2.92	1.38	2.81	5.71

Determine the variance, standard deviation and relative variability about the mean. Then group the data into seven intervals and repeat the calculations.

6.9 Eliminating Outliers

ALL STATISTICS WHICH involve summing the observations are affected by the presence of outliers. The outliers on the low end might balance those on the high end when talking about the mean, but with the deviations this cannot happen because either they have an absolute value taken, or in the case of variation and variance are squared (which makes their impact even worse). As a result, when one is confronted with the average and median absolute variability, the average and median absolute deviation, the variation, standard deviance, variance or standard deviation, it is even more critical to be able to handle outliers. In our discussion of the mean we discovered that there are two methods for eliminating outliers, trimming and Winsorizing. These methods can be used in the same way for all of these methods.

Trimming

AS WE HAVE SEEN before, trimming can be done in two ways. **Absolute trimming** cuts off a fixed number from each end. **Percentage trimming** cuts off a fixed percentage from each end. We have also noted that the quantiles of percentage trimming can be done using the traditional sample method or the restricted proper method as these are the methods which treat the sample as if it were a population. Because trimming treats the sample as a population, the denominator is the number of items being summed, and not one less than this, when trimming the variation, standard deviance, variance or standard deviation.

6.9. ELIMINATING OUTLIERS

Example 1:
Determine the 10%-trimmed average and median absolute variability for the dice data.

Answer:
The tally table for the data is

$$
\begin{array}{llllll}
2\colon 1 & 4\colon 4 & 6\colon 7 & 8\colon 7 & 10\colon 5 & 12\colon 1 \\
3\colon 5 & 5\colon 6 & 7\colon 8 & 9\colon 5 & 11\colon 1 &
\end{array}
$$

By the traditional sample method,

$$\#q(0.1) = 0.1 \cdot 50 = 5 \text{ and } s = 6;$$

also,

$$\#q(0.9) = 0.9 \cdot 50 = 45 = t.$$

Thus, only the $45 - 6 + 1 = 40$ observations stretching between these limits are used to calculate differences. This reduces the frequency table to

$$
\begin{array}{llll}
3\colon 1 & 5\colon 6 & 7\colon 8 & 9\colon 5 \\
4\colon 4 & 6\colon 7 & 8\colon 7 & 10\colon 2
\end{array}
$$

The difference table based upon this reduced difference table is

	4	5	6	7	8	9	10
3	1	2	3	4	5	6	7
	4	6	7	8	7	5	2
4		1	2	3	4	5	6
		24	28	32	28	20	8
5̄			1	2	3	4	5
			42	48	42	30	12
6̄				1	2	3	4
				56	49	35	14
7̄					1	2	3
					56	40	16
8̄						1	2
						35	14
9̄							1
							10

There are $40 \cdot 39/2 = 780$ differences instead of the original 1225 differences when all of the data were used. The tally table of absolute differences is

$$
\begin{array}{llll}
0\colon 102 & 2\colon 185 & 4\colon 80 & 6\colon 13 \\
1\colon 227 & 3\colon 132 & 5\colon 39 & 7\colon\ 2
\end{array}
$$

The mean of the above is the trimmed statistic sought; hence,

$$\text{aav} = \frac{1 \cdot 227 + \cdots + 7 \cdot 2}{780} = \frac{1600}{780} = \frac{80}{39} \approx 2.0513.$$

Compare this with the value $135/49 \approx 2.7751$, from Example 2, §6.2, starting on page 293, a reduction of more than 25%.

As the percentiles were calculated using the traditional sample method, we use this method to determine the median

$$\text{mav} = \frac{d_{(390)} + d_{(391)}}{2} = \frac{2+2}{2} = 2,$$

exactly what we got before.

By the restricted proper method, both $x_{\min} = 2$ and $x_{\max} = 12$ are unique, so

$$\#q(0.1) = \frac{9 \cdot 1 + 50}{10} = 5.9 \text{ and } s = 6.$$

Also,

$$\#q(0.9) = \frac{1 + 9 \cdot 50}{10} = 45.1 \text{ and } t = 45.$$

We conclude aav $= 80/39 \approx 2.0513$, as before.

Using the restricted proper method for the percentiles means we should use it for the median, as well.

$$\#(d_{\min}) = \frac{1 + 102}{2} = 51.5 \text{ and } \#(d_{\max}) = \frac{779 + 780}{2} = 779.5.$$

We conclude

$$\#(\text{mav}) = \frac{51.5 + 779.5}{2} = 415.5.$$

6.9. ELIMINATING OUTLIERS

Now, $d_{(415)} = 2$ and $\#(2) = (330 + 514)/2 = 422 > 415.2$; thence, $y_2 = 2$, $m_2 = 422$, $y_1 = 1$ and $m_1 = (103 + 329)/2 = 216$. By interpolation,

$$\text{mav} = \frac{1 \cdot (422 - 415.5) + 2 \cdot (415.5 - 216)}{422 - 216}$$
$$= \frac{6.5 + 399}{206} = \frac{811}{412} \approx 1.9684.$$

Compare this with $1155/446 \approx 2.5897$, a reduction of almost 24%.

Example 2:

Determine the 2-trimmed average and median absolute variability for the two-success data.

Answer:

The original sorted data are

2	2	2	2	2	2	2	2	2	3
3	3	3	3	3	3	3	3	3	3
3	3	3	3	3	4	4	4	4	4
4	4	4	4	4	4	4	4	4	4
5	5	5	5	5	5	5	5	5	5
5	5	5	5	5	5	5	6	6	6
6	6	6	6	6	6	7	7	7	7
8	8	8	8	9	9	9	9	9	10
10	10	10	10	10	11	11	11	11	12
12	14	14	14	14	14	15	16	17	28

There were 100 observations, originally. The 2-trimmed average absolute variability will ignore the first two and last two observations to reduce the number of observations down to 96. This makes the trimmed frequency table for this data:

2: 7   4: 15   6: 9   8: 4   10: 6   12: 2   15: 1
3: 16   5: 17   7: 4   9: 5   11: 4   14: 5   16: 1

which produces the following absolute difference table: (the zero differences have not been shown in order to reduce the space)

	3	4	5	6	7	8	9	10	11	12	14	15	16
2	1 112	2 105	3 119	4 63	5 28	6 28	7 35	8 42	9 28	10 14	12 35	13 7	14 7
3		1 240	2 272	3 144	4 64	5 64	6 80	7 96	8 64	9 32	11 80	12 16	13 16
4			1 255	2 135	3 60	4 60	5 75	6 90	7 60	8 30	10 75	11 15	12 15
5				1 153	2 68	3 68	4 85	5 102	6 68	7 34	9 85	10 17	11 17
6					1 36	2 36	3 45	4 54	5 36	6 18	8 45	9 9	10 9
7						1 16	2 20	3 24	4 16	5 8	7 20	8 4	9 4
8							1 20	2 24	3 16	4 8	6 20	7 4	8 4
9								1 30	2 20	3 10	5 25	6 5	7 5
10									1 24	2 12	4 30	5 6	6 6
11										1 8	3 20	4 4	5 4
12											2 10	3 2	4 2
14												1 5	2 5
15													1 1

There are $96 \cdot 95/2 = 4560$ differences in total and we get the following absolute difference tally table.

0: 472	3: 508	6: 315	9: 158	12: 66
1: 900	4: 386	7: 254	10: 115	13: 23
2: 707	5: 348	8: 189	11: 112	14: 7

The mean of these absolute differences is the statistic we seek.

$$\text{aav} = \frac{1 \cdot 900 + \cdots + 14 \cdot 7}{4560} = \frac{15015}{4560} = \frac{1001}{304} \approx 3.2928.$$

6.9. ELIMINATING OUTLIERS

Compare this result to the result of Example 1, §6.2, which started on page 291 and the variability has gone down from $3554/825 \approx 4.3079$ to its present value, a more than 23.5% reduction. This gives you an idea of how much outliers can affect data variability.

Notice how eliminating 4 observations reduced the number of differences from 4950 to 4560. This is because you are eliminating all of the interactions involving the eliminated observations, as well.

The median-of-medians and traditional sample method give

$$\text{mav} = \frac{d_{(2280)} + d_{(2281)}}{2} = \frac{3+3}{2} = 3,$$

exactly as before. The restricted proper method notes that

$$\#(d_{\min}) = \frac{1+472}{2} = 236.5 \text{ and } \#(d_{\max}) = frac{4554+4560}{2} = 4557.$$

Therefore,

$$\#(\text{mav}) = \frac{236.5 + 4557}{2} = 2396.75.$$

Now, $d_{(2396)} = 3$ and $\#(3) = (2080 + 2587)/2 = 2333.5 < 2396.75$. Therefore, $y_1 = 3$, $m_1 = 2333.5$, $y_2 = 4$ and $m_2 = (2587 + 2972)/2 = 2779.5$. By interpolation,

$$\text{mav} = \frac{3 \cdot (2779.5 - 2396.75) + 4 \cdot (2396.75 - 2333.5)}{2779.5 - 2333.5}$$
$$= \frac{1148.25 + 253}{446} = \frac{5605}{1764} \approx 3.1418.$$

Compare this with $3179/951 \approx 3.3428$ which is only about a 6% decrease in magnitude.

Trimming r observations from each end for the other six statistics is similar — just think of it as reducing the size of the sample

— except with those based upon the root mean square average are treated as if they were each acting on a census, not a sample, so there is no subtracting one in the denominator.

The other way of trimming involves getting rid of a fixed percentage of values from each end. In many cases, more actual observations are trimmed from one end than the other, but the number trimmed from each end will never exceed one.

Winsorizing

SOMETIMES THE EFFECT of trimming is too much. The other way of eliminating outliers is to Winsorize the data. Basically, you keep the same number of data, but the end data are replaced, either by the last value kept, or by the quantile corresponding to the given percentile. Thus, there are again two ways to perform this operation.

Winsorizing has the effect of dulling, but not eliminating, the outside information. As a rule-of-thumb, if the trimmed statistic differs by more than 15% from the untrimmed value then one should probably consider Winsorizing rather than trimming — especially if there are multiple outliers in the same direction.

An r-Winsorized statistic replaces the first r values by $x_{(r+1)}$ and the last r values by $x_{(n-r)}$. A $100p\%$-Winsorized statistic has those values up to $x_{(\lfloor \#q(p) \rfloor)}$ replaced by $q(p)$ and those values starting at $x_{(\lfloor \#q(1-p) \rfloor + 1)}$ and going up to $x_{(n)}$ replaced by $q(1-p)$.

As we have seen, variability statistics are based upon all possible differences among the data. Deviations, however, are based upon differences about a fixed point. If the point is the origin then we get a central deviation. If the point is the midrange then we get a midpoint deviation. The other two commonly used fixed points

6.9. ELIMINATING OUTLIERS

are the median (done either by the traditional sample or restricted proper method) and the mean.

Calculating a deviation always follows the same steps: 1) calculate the value of the fixed point, 2) subtract the fixed point from each observation, and 3) perform whatever operations are needed to calculate the statistic from these differences. This could involve either taking the absolute value of the difference or squaring that difference. It could then involve either a simple mean, or the root of that mean on the modified differences.

Unlike trimming, Winsorizing does not affect the formula for variation, variance, standard deviance or standard deviation; that is, on a sample which is not a census one must reduce the denominator by one.

Example 3:
Determine the 3-Winsorized variation and standard deviance for the random gamma data.

Answer:

The modified sorted data is below, each datum from $x_{(1)}$ to $x_{(3)}$ has been replaced by $x_{(4)}$ (actually only x_1 was affected) and each datum from $x_{(23)}$ to $x_{(25)}$ has been replaced by $x_{(22)}$ (actually only the last two were affected). [We conclude from this that the 2-Winsorized and 3-Winsorized statistics for this data will all be the same no matter what statistic is being calculated from them.]

5	5	5	5	5
8	9	9	10	12
13	13	13	14	14
15	17	17	18	22
22	24	24	24	24

This give the following modified tally table

CHAPTER 6. DATA VARIABILITY

5: 5	9: 2	12: 1	14: 2	17: 2	22: 2
8: 1	10: 1	13: 3	15: 1	18: 1	24: 4

By the traditional sample method (and equivalently the median-of-medians method), the median is

$$x_{(13)} = 13.$$

Subtracting this from each value, squaring and combining like values gives

0: 3	9: 1	64: 5
1: 3	16: 4	81: 2
4: 1	25: 2	121: 4

This makes the variation

$$V = \frac{1 \cdot 3 + \cdots + 121 \cdot 4}{25 - 1} = \frac{1096}{24} = \frac{137}{3} \approx 45.66666667$$

and thus the standard deviance is

$$s_d \approx 6.757712.$$

Example 4:
Determine the 15%-Winsorized variance and standard deviation for the two-success data.

Answer:
The frequency table for this data is

2: 9	4: 15	6: 9	8: 4	10: 6	12: 2	15: 1	17: 1
3: 16	5: 17	7: 4	9: 5	11: 4	14: 5	16: 1	28: 1

By the traditional sample method,

$$\#q(0.15) = 0.15 \cdot 100 = 15 \text{ and } s = 16; \text{ so, } q(0.15) = \frac{3+3}{2} = 3$$

6.9. ELIMINATING OUTLIERS

as $x_{(15)} = x_{(16)} = 3$. Furthermore,

$$\#q(0.15) = 0.85 \cdot 100 = 85 = t \text{ and } q(0.85) = \frac{10 + 11}{2} = 10.5$$

as $x_{(85)} = 10$ and $x_{(86)} = 11$.

These calculations result in the following frequency table

$$
\begin{array}{lll}
3\colon 25 & 6\colon 9 & 9\colon 5 \\
4\colon 15 & 7\colon 4 & 10\colon 6 \\
5\colon 17 & 8\colon 4 & 10.5\colon 15
\end{array}
$$

From the above,

$$\sum_{i=1}^{100} x_i = 3 \cdot 25 + \cdots + 10.5 \cdot 15 = 596.5.$$

In addition,

$$\sum_{i=1}^{100} x_i^2 = 3^2 \cdot 25 + \cdots + 10.5^2 \cdot 15 = 4324.75.$$

The calculating formula for the variance is

$$V = \frac{100 \cdot 4324.75 - (596.5)^2}{100 \cdot 99} = \frac{76662.75}{9900}$$
$$= \frac{102217}{13200} \approx 7.74371212.$$

We conclude that the standard deviation is

$$s \approx 2.782753.$$

Exercises 6.9

1. One day while walking through a field you notice there seem to be a lot of stones lying on the ground. Curious about what the average size of a stone is in the field you randomly collect 30 stones. You weigh each stone and record its weight in grams. The following table shows the raw data.

110	126	75	110	126	94
90	112	111	139	131	92
102	103	96	75	67	101
104	111	100	65	92	93
106	104	93	115	109	95

   Determine the 2-trimmed and 15%-Winsorized statistics for the average and median absolute variability, the average and median absolute deviation about the median, the variation and standard deviance and the variance and standard deviation for this data.

2. The scores, in the order graded, on a recent midterm examination in Statistics, a class with 30 students, are given below.

70	80	99	98	85	89
87	79	83	38	69	70
60	69	78	40	75	56
70	51	99	69	95	86
57	53	47	50	55	81

   Determine the 2-trimmed and 15%-Winsorized statistics for the average and median absolute variability, the average and median absolute deviation about the median, the variation and standard deviance and the variance and standard deviation for this data.

3. Below is a sample of the monthly salary for 30 randomly selected people in mid-management in thousands of dollars.

3.08	3.18	3.13	7.41	3.73	3.05
4.17	3.58	3.36	3.27	3.36	4.74
3.32	3.61	4.26	3.02	3.45	4.06
6.73	4.72	3.13	3.15	3.70	3.59
3.12	3.61	5.03	3.53	3.20	3.32

6.9. ELIMINATING OUTLIERS

Determine the 2-trimmed and 15%-Winsorized statistics for the average and median absolute variability, the average and median absolute deviation about the median, the variation and standard deviance and the variance and standard deviation for this data.

4. A pair of dice, one red and one green, is rolled in the following manner. On the first toss both dice are tossed. Then the red die is tossed while the green die keeps its value to get the second sum. Then the green die is tossed while the red die keeps its value. You continue tossing alternately the red then the green, die and generate the following table.

8	11	8	6	10	9	9
7	2	5	10	11	8	5
3	6	9	7	7	9	10
7	3	3	7	9	8	7
5	4	8	11	11	10	10

Determine the 2-trimmed and 15%-Winsorized statistics for the average and median absolute variability, the average and median absolute deviation about the median, the variation and standard deviance and the variance and standard deviation for this data.

5. The following random sample of 72 observations comes from a gamma distribution, so $\mu_{min} = 0$ and μ_{max} has no upper limit. The raw data are

2.15	1.92	1.92	2.20	3.67	4.55	3.32	2.71	4.87
2.09	1.13	1.09	2.12	4.89	3.03	3.17	4.74	4.44
4.61	2.78	2.97	2.78	2.88	3.66	2.55	4.40	3.54
2.18	3.43	3.84	4.60	1.76	2.85	2.27	1.38	1.24
4.22	1.67	4.51	3.50	6.69	5.92	6.42	0.95	4.57
3.10	3.26	2.95	2.45	1.95	2.06	3.10	2.03	4.98
0.86	2.61	1.92	2.78	1.01	2.64	3.14	0.86	3.68
2.71	3.06	4.33	4.63	2.47	2.92	1.38	2.81	5.71

Determine the 2-trimmed and 15%-Winsorized statistics for the average and median absolute variability, the average and median absolute deviation about the median, the variation and standard deviance and the variance and standard deviation for this data.

6.10 Standard Z-Scores

THE RELATIVE variability is one way of measuring whether the variability of the data is large or small compared with the size of the data. Another method is to compare it to the standard normal curve — the same curve that students talk about when they ask whether I grade "on the curve," which I do not. There is the famous story of the class that begged the professor to curve the grade. The professor gave them an easy test to demonstrate why that was a bad idea. The average on the test was 92% — which became the C — the two people who scored 85% failed. The pass and fail divisions were calculated using Z-scores.

A **z-score** for an observation x, denoted $z(x; \mu, \sigma)$, subtracts the mean (μ) from x and divides the result by the standard deviation (σ).

$$z(x; \mu, \sigma) = \frac{x - \mu}{\sigma}.$$

Example 1:
Determine the standard z-scores for the following measurements assuming $\mu = 10$ and $\sigma = 4$.

$$-2 \quad 4 \quad 8 \quad 14 \quad 22$$

Answer:
When $x = -2$,

$$z(-2; 10, 4) = \frac{-2 - 10}{4} = -\frac{12}{4} = -3.$$

For $x = 4$,

$$z(4; 10, 4) = \frac{4 - 10}{4} = -\frac{6}{4} = -1.5.$$

6.10. STANDARD Z-SCORES

If $x = 8$ then
$$z(8; 10, 4) = \frac{8 - 10}{4} = -\frac{2}{4} = -0.5.$$

Given $x = 14$ and the other values above,
$$z(14; 10, 4) = \frac{14 - 10}{4} = \frac{4}{4} = 1.$$

If $x = 22$ then
$$z(22; 10, 4) = \frac{22 - 10}{4} = \frac{12}{4} = 3.$$

Knowing the standard z-score, the mean and standard deviation it is possible to determine the measurement.

$$x = \mu + z \cdot \sigma$$

Example 2:
IQ scores are normed to ensure $\mu = 100$ and $\sigma = 16$. A genius is a person whose IQ score is at least 2.5 standard deviations above the mean. What is the minimum IQ score of a genius?

Answer:
   We have been directly told $\mu = 100$ and $\sigma = 16$ and indirectly told that $z \geq 2.5$. Thus,
$$\text{IQ score} = 100 + 2.5 \cdot 16 = 100 + 40 = 140.$$

Example 3:
Determine the z-scores for the observations of the dice roll data. Round your final answers to two decimal places.

Answer:
In Example 3, §6.8, starting on page 344, we determined
$$\sum_{i=1}^{50} x_i = 335 \text{ and } \sum_{i=1}^{50} x_i^2 = 2525.$$
This means
$$\mu = \bar{x} = \frac{335}{50} = \frac{67}{10} = 6.7$$
and
$$\sigma = s = \sqrt{\frac{50 \cdot 2525 - 335^2}{50 \cdot 49}} = \frac{\sqrt{1122}}{14} \approx 2.392591.$$
The observations are the 11 integers starting at 2 and going up to 12.
For $x = 2$,
$$z \approx \frac{2 - 6.7}{2.392591} \approx -1.96.$$
When $x = 3$,
$$z \approx \frac{3 - 6.7}{2.392591} \approx -1.55.$$
Given $x = 4$,
$$z \approx \frac{4 - 6.7}{2.392591} \approx -1.13.$$
If $x = 5$ then
$$z \approx \frac{5 - 6.7}{2.392591} \approx -0.71.$$
For $x = 6$,
$$z \approx \frac{6 - 6.7}{2.392591} \approx -0.29.$$
When $x = 7$,
$$z \approx \frac{7 - 6.7}{2.392591} \approx 0.13.$$
Given $x = 8$,
$$z \approx \frac{8 - 6.7}{2.392591} \approx 0.54.$$

6.10. STANDARD Z-SCORES

If $x = 9$ then
$$z \approx \frac{9 - 6.7}{2.392591} \approx 0.96.$$

For $x = 10$,
$$z \approx \frac{10 - 6.7}{2.392591} \approx 1.38.$$

When $x = 11$,
$$z \approx \frac{11 - 6.7}{2.392591} \approx 1.80.$$

Given $x = 12$,
$$z \approx \frac{12 - 6.7}{2.392581} \approx 2.22.$$

If the data are normal then Table 6.1 gives an idea of the amount of data which is k standard deviation from the mean; that is, it gives the percentage of the data which should be in the interval $\mu - k \cdot \sigma$ to $\mu + k \cdot \sigma$.

Table 6.1: Percentage within $\mu \pm k \cdot \sigma$.

k	Percentage
1	68.2689492137
2	95.4499736104
3	99.7300203937
4	99.9936657516
5	99.9999426697

Example 4:
Consider the two-success data. How many tries is 4 standard deviations above the mean? Give your answer as the least integer greater than or equal to what you calculated.

Answer:

The frequency table for this data is

2: 9	4: 15	6: 9	8: 4	10: 6	12: 2	15: 1	17: 1
3: 16	5: 17	7: 4	9: 5	11: 4	14: 5	16: 1	28: 1

Therefrom,
$$\sum_{i=1}^{100} x_i = 644 \text{ and } \sum_{i=1}^{100} x_i^2 = 5932.$$

The mean is
$$\mu = \bar{x} = \frac{644}{100} = \frac{161}{25} = 6.44$$

and the variance is
$$\sigma^2 = s^2 = V = \frac{100 \cdot 5932 - 644^2}{100 \cdot 99} = \frac{178464}{9900} = \frac{1352}{75}.$$

We conclude the standard deviation is
$$\sigma = s = \sqrt{\frac{1352}{75}} = \frac{26\sqrt{6}}{15} \approx 4.245782.$$

The point 4 standard deviations above the mean is
$$x \approx 6.44 + 4 \cdot 4.245782 \approx 23.42.$$

The answer is $x = 24$.

To get a feel for this, if this data followed the normal curve (it isn't) then the probability of being 4 standard deviation above the mean is about 0.0000316712 (less than 1 chance in 31575 tries). Now, $x = 28$ is even higher! This is clearly an unusual occurrence.

6.10. STANDARD Z-SCORES

Exercises 6.10

1. One day while walking through a field you notice there seem to be a lot of stones lying on the ground. Curious about what the average size of a stone is in the field you randomly collect 30 stones. You weigh each stone and record its weight in grams. The following table shows the raw data.

110	126	75	110	126	94
90	112	111	139	131	92
102	103	96	75	67	101
104	111	100	65	92	93
106	104	93	115	109	95

   Calculate the percentage of the data within 2 standard deviations and compare it to Table 6.1. Do you think this data might be normal?

2. The scores, in the order graded, on a recent midterm examination in Statistics, a class with 30 students, are given below.

70	80	99	98	85	89
87	79	83	38	69	70
60	69	78	40	75	56
70	51	99	69	95	86
57	53	47	50	55	81

   Calculate the percentage of the data within 2 standard deviations and compare it to Table 6.1. Do you think this data might be normal?

3. Below is a sample of the monthly salary for 30 randomly selected people in mid-management in thousands of dollars.

3.08	3.18	3.13	7.41	3.73	3.05
4.17	3.58	3.36	3.27	3.36	4.74
3.32	3.61	4.26	3.02	3.45	4.06
6.73	4.72	3.13	3.15	3.70	3.59
3.12	3.61	5.03	3.53	3.20	3.32

What are the standard z-scores for the data extremes? Give your final results rounded to two decimal places.

4. A pair of dice, one red and one green, is rolled in the following manner. On the first toss both dice are tossed. Then the red die is tossed while the green die keeps its value to get the second sum. Then the green die is tossed while the red die keeps its value. You continue tossing alternately the red then the green, die and generate the following table.

8	11	8	6	10	9	9
7	2	5	10	11	8	5
3	6	9	7	7	9	10
7	3	3	7	9	8	7
5	4	8	11	11	10	10

Determine the standard z-scores for all distinct observations and give the final result to two decimal places. How do your answers compare with Example 3 above?

5. The following random sample of 72 observations comes from a gamma distribution, so $\mu_{min} = 0$ and μ_{max} has no upper limit. The raw data are

2.15	1.92	1.92	2.20	3.67	4.55	3.32	2.71	4.87
2.09	1.13	1.09	2.12	4.89	3.03	3.17	4.74	4.44
4.61	2.78	2.97	2.78	2.88	3.66	2.55	4.40	3.54
2.18	3.43	3.84	4.60	1.76	2.85	2.27	1.38	1.24
4.22	1.67	4.51	3.50	6.69	5.92	6.42	0.95	4.57
3.10	3.26	2.95	2.45	1.95	2.06	3.10	2.03	4.98
0.86	2.61	1.92	2.78	1.01	2.64	3.14	0.86	3.68
2.71	3.06	4.33	4.63	2.47	2.92	1.38	2.81	5.71

Determine the value, rounded to two decimal places, for the measurement which lies 3.5 standard deviations above the mean. Compare your measurement to the sample maximum.

Chapter 7

Skewness

The word SKEW *comes from the word* skuen *which means "to shy away from" or "to avoid." Skewness, is then the property which measures by how much the data tend to avoid either high or low values. Many books use the phrase "skewed to" but that just does not make sense when you realize what it means. It should be "skewed from" and that is how we will word things.*

7.1 The Tilt

PERHAPS THE SIMPLEST of all skewness statistics, the **tilt**, denoted τ, is the number of data below the midrange minus the number of data above it, divided by two less than the number of data not equal to x_{mid}.

We calculate the tilt as follows: Assume there are n data in the sample and sort them from smallest to largest. Let j be the largest integer such that $x_{(j)} < x_{\text{mid}}$ and k the smallest integer such that

$x_{(k)} > x_{\text{mid}}$. Set N to be the number of data equal to x_{mid}, if there are any. The tilt is defined as

$$\tau = \frac{j + k - (n+1)}{n - 2 - N}.$$

The above always produces a statistic in the interval $[-1, 1]$.

To see why this works we consider that there must be at least one on each side of the midrange as $x_{\min} < x_{\text{mid}} < x_{\max}$. The maximum number of values which could tip the scale must then be $n - 2$. Now all values which equal x_{mid} cannot possibly tip the scale either direction. Thus, the denominator serves as a normalizing quantity to ensure $-1 \le \tau \le 1$.

There are j values below x_{mid} by the definition of j. There are $n - k + 1$ above x_{mid}. This latter can be seen by numbering the indices

$$x_{(k+0)}, x_{(k+1)}, \ldots, x_{(k+[n-k])}.$$

Finally we perform the subtraction to get

$$j - (n - k + 1) = j - n + k - 1 = j + k - (n+1),$$

which is the numerator.

Example 1:
Determine the tilt for the oyster data.

Answer:
The sorted data are

5.76	6.10	6.36	6.37	6.42	6.53	6.63	6.67	6.74	6.91
6.91	6.97	7.03	7.04	7.37	7.40	7.71	7.97	8.02	8.04
8.12	8.14	8.42	8.53	8.61	8.62	9.01	9.20	9.21	9.39
9.47	9.53	9.55	9.98	10.24	10.50	10.62	10.63	10.63	10.81
10.81	10.86	11.02	11.09	11.17	11.31	11.59	11.97	12.07	13.18

7.1. THE TILT

From which we see

$$x_{\min} = \frac{5.76 + 13.18}{2} = 9.47.$$

The largest integer j such that $x_{(j)} < 9.47$ is $j = 30$ as $x_{(30)} = 9.39$ while $x_{(31)} = 9.47 = x_{\mathrm{mid}}$.

The smallest integer k such that $x_{(k)} > 9.47$ is $k = 32$ as $x_{(32)} = 9.53$ while $x_{(31)} = 9.47 = x_{\mathrm{mid}}$.

There is exactly one datum $x_{(31)} = x_{\mathrm{mid}}$ so $N = 1$ and there are $n = 50$ data altogether.

Therefore,

$$\tau = \frac{30 + 32 - 51}{48 - 1} = \frac{11}{47} \approx 0.234$$

Table 7.1: Skewing Terminology

$\tau < 0$	negatively skewed skewed from the left	more higher values than lower values
$\tau = 0$	zero skewing in balance	equal numbers above and below midrange
$\tau > 0$	positively skewed skewed from the right	more lower values than higher values

Whenever $\tau > 0$, there are more data below the midrange than above it. This means the data are skewed from the higher values; that is, **skewed from the right**. Therefore, if data are **positively skewed**, they are skewed from the right.

Example 2:
Determine the tilt for the C++ midterm data.

Answer:

The sorted data are

| 46 | 55 | 62 | 67 | 68 | 68 | 72 | 74 | 75 | 76 | 77 |
| 79 | 79 | 80 | 80 | 81 | 82 | 86 | 88 | 91 | 94 | 95 |

There are 22 data and we can see that

$$x_{\text{mid}} = \frac{46 + 95}{2} = 70.5.$$

There are 6 data less than 70.5 and 16 greater than 70.5 — none are equal to it. Therefore,

$$\tau = \frac{6 - 16}{20} = -\frac{1}{2} = -0.5.$$

Whenever $\tau < 0$, there are more data above the midrange than below it. This means the data are skewed from the lower values; that is, **skewed from the left**. Therefore, if data are **negatively skewed**, they are skewed from the left.

If the data are neither skewed from the right nor from the left then $\tau = 0$ and the data are **in balance**. We sometimes say that they are **symmetric about the midrange**.

These facts about skewing are summarized in Table 7.1.

Tilt Across the Mean

ALTHOUGH THE tilt is normally calculated from the midrange, it does not need to be. It can also be calculated across the mean. The calculation of **tilt across the mean** is identical to the normal tilt, only the value where you calculate j, k and N differ. We will use τ_M for the tilt calculated from the mean. The main thing to realize is that the tilt across the mean and the normal tilt (across the midrange) may give contradictory results.

7.1. THE TILT

The tilt across the mean is not significant when $|\tau_M| < 0.125$. It is borderline when $0.125 \leq |\tau_M| \leq 0.25$ and it is significant when $|\tau_M| > 0.25$. This is summarized in Table 7.2. The sign is interpreted the same way as the tilt; however, if the result is not significant then we consider the data to be balanced (the skewing is essentially zero).

Table 7.2: Significance of tilt across mean.

$\|\tau_M\| < 0.125$	not significant
$0.125 \leq \|\tau_M\| \leq 0.25$	borderline
$\|\tau_M\| > 0.25$	significant

Example 3:
Use the tilt across the mean to determine whether the oyster data are significantly skewed. If they are, then what is the direction of skewing?

Answer:
The mean for the data are
$$x = \frac{5.76 + \cdots + 13.18}{50} = \frac{443.23}{50} = \frac{44323}{5000} = 8.8646.$$

There are 26 data below 8.8646 and 24 above, with none equal. We conclude
$$\tau_M = \frac{26 - 24}{50 - 2} = \frac{2}{48} = \frac{1}{24} \approx 0.041667 < 0.125.$$
The data are not significantly skewed across the mean.
Notice: $j = 26$, $k = 27$ and $N = 0$; thus,
$$\tau_M = \frac{26 + 27 - (50 + 1)}{50 - 2 - 0} = \frac{2}{48} = \frac{1}{24}.$$
This verifies that both ways of calculating the formula are exactly the same.

When skewing about the mean is not significant, but there is skewing about the midrange, then we say the data are **shifted**. If $\tau < 0$ then the data are shifted right. If $\tau > 0$ then the data are shifted left. Remember: **The shift is in the opposite direction from the skewing.** In the above example the data are shifted left.

Example 4:
Determine whether the C++ midterm data are skewed or shifted. If either is true, then what is the direction?
Answer:
The mean of the midterm data is
$$\bar{x} = \frac{46 + \cdots + 95}{22} = \frac{1674}{22} = \frac{837}{11} \approx 76.09.$$

There are 10 scores below the mean and 12 above, with none equal; therefore,
$$\tau_M = \frac{10 - 12}{22 - 2} = \frac{\text{-}2}{20} = \text{-}0.1.$$
Now, $0.1 < 0.125$, so the data are not significantly skewed.

Because $\tau < 0$, we conclude the data are shifted right. This does make sense, because there are more higher values than lower values relative to the midrange. The fact that the skewing is in the same direction just adds to the shifted effect as it is not significantly skewed.

Example 5:
Determine whether the random gamma data are skewed with respect to the mean. Also, determine the tilt for this data.
Answer:
We need to determine both τ and τ_M for this data. The tally table for the data is

7.1. THE TILT

4: 1	9: 2	13: 3	17: 2	24: 2
5: 4	10: 1	14: 2	18: 1	27: 1
8: 1	12: 1	15: 1	22: 2	31: 1

The midrange is

$$x_{\text{mid}} = \frac{4+31}{2} = \frac{35}{2} = 17.5.$$

The largest integer j such that $x_{(j)} < 17.5$ is $j = 18$. The smallest integer k such that $x_{(k)} > 17.5$ is $k = 19$. There are 25 observations with none equal to x_{mid}. Therefore,

$$\tau = \frac{18 + 19 - (25+1)}{25 - 2 - 0} = \frac{11}{23} \approx 0.478261.$$

The mean of the data is

$$x = \frac{4 \cdot 1 + \cdots + 31 \cdot 1}{25} = \frac{356}{25} = 14.24.$$

There are 15 observations less that 14.24 and 10 greater with none equal. Therefore,

$$\tau_M = \frac{15 - 10}{25 - 2} = \frac{5}{23} = 0.217391 \leq 0.25.$$

There is borderline skewing from the right.

Overall, the data are shifted left but not significantly skewed.

Exercises 7.1

1. One day while walking through a field you notice there seem to be a lot of stones lying on the ground. Curious about what the average

size of a stone is in the field you randomly collect 30 stones. You weigh each stone and record its weight in grams. The following table shows the raw data.

110	126	75	110	126	94
90	112	111	139	131	92
102	103	96	75	67	101
104	111	100	65	92	93
106	104	93	115	109	95

Determine the tilt for this data and identify the type of skewing as listed in Table 7.1. Then determine the tilt across the mean to determine the same thing. If the tilt across the mean is not significant then are the data shifted? If so, what way?

2. The scores, in the order graded, on a recent midterm examination in Statistics, a class with 30 students, are given below.

70	80	99	98	85	89
87	79	83	38	69	70
60	69	78	40	75	56
70	51	99	69	95	86
57	53	47	50	55	81

Determine the tilt for this data and identify the type of skewing as listed in Table 7.1. Then determine the tilt across the mean to determine the same thing. If the tilt across the mean is not significant then are the data shifted? If so, what way?

3. Below is a sample of the monthly salary for 30 randomly selected people in mid-management in thousands of dollars.

3.08	3.18	3.13	7.41	3.73	3.05
4.17	3.58	3.36	3.27	3.36	4.74
3.32	3.61	4.26	3.02	3.45	4.06
6.73	4.72	3.13	3.15	3.70	3.59
3.12	3.61	5.03	3.53	3.20	3.32

Determine the tilt for this data and identify the type of skewing as listed in Table 7.1. Then determine the tilt across the mean to

7.1. THE TILT

determine the same thing. If the tilt across the mean is not significant then are the data shifted? If so, what way?

4. A pair of dice, one red and one green, is rolled in the following manner. On the first toss both dice are tossed. Then the red die is tossed while the green die keeps its value to get the second sum. Then the green die is tossed while the red die keeps its value. You continue tossing alternately the red then the green, die and generate the following table.

8	11	8	6	10	9	9
7	2	5	10	11	8	5
3	6	9	7	7	9	10
7	3	3	7	9	8	7
5	4	8	11	11	10	10

Determine the tilt for this data and identify the type of skewing as listed in Table 7.1. Then determine the tilt across the mean to determine the same thing. If the tilt across the mean is not significant then are the data shifted? If so, what way?

5. The following random sample of 72 observations comes from a gamma distribution, so $\mu_{min} = 0$ and μ_{max} has no upper limit. The raw data are

2.15	1.92	1.92	2.20	3.67	4.55	3.32	2.71	4.87
2.09	1.13	1.09	2.12	4.89	3.03	3.17	4.74	4.44
4.61	2.78	2.97	2.78	2.88	3.66	2.55	4.40	3.54
2.18	3.43	3.84	4.60	1.76	2.85	2.27	1.38	1.24
4.22	1.67	4.51	3.50	6.69	5.92	6.42	0.95	4.57
3.10	3.26	2.95	2.45	1.95	2.06	3.10	2.03	4.98
0.86	2.61	1.92	2.78	1.01	2.64	3.14	0.86	3.68
2.71	3.06	4.33	4.63	2.47	2.92	1.38	2.81	5.71

Determine the tilt for this data and identify the type of skewing as listed in Table 7.1. Then determine the tilt across the mean to determine the same thing. If the tilt across the mean is not significant then are the data shifted? If so, what way?

7.2 Relative Difference of the Middles

WHEN BOTH THE mean and the median of the data are known then there is a second simple statistic for skewness. The **relative difference of the middles**, denoted Δ_m, is defined as

$$\Delta_m = \frac{\bar{x} - M}{M}.$$

Just as with the tilt, the sign of Δ tells whether the data have positive, negative or zero skewing. Table 7.3 summarizes the level of significance for this statistic. The sign of the statistic is interpreted the same as for the tilt: positive skewing avoids higher values; it has fewer observations above the mean than below it. Negative skewing avoids lower values; it has fewer observations below the mean than above it. Zero skewing means the mean and median are not significantly different and hence can be considered equal with the data in balance.

Table 7.3: Significance of relative difference of the middles.

$\lvert \Delta_m \rvert < 0.05$	not significant
$0.05 \leq \lvert \Delta_m \rvert \leq 0.1$	borderline
$\lvert \Delta_m \rvert > 0.1$	significant

Example 1:
Determine the relative difference of the middles and use it to classify whether the two-success data are significantly skewed. If so, from which direction are they skewed?
Answer:

7.2. RELATIVE DIFFERENCE OF THE MIDDLES

The mean was calculated in Example 1, §5.7, page 229 to be $\bar{x} = 6.44$.

The traditional sample method calculated the median to be $M = 5$ in Example 1, §5.3, page 194. This would make,

$$\Delta_m = \frac{6.44 - 5}{5} = \frac{36}{125} = 0.288 > 0.1$$

and we would declare the data to be significantly skewed from the right — there are more lower values than higher values.

The same example made $M = 137/26$ by the restricted proper method. This would mean

$$\Delta_m = \frac{161/25 - 137/26}{137/26} = \frac{761}{3425} \approx 0.2221898 > 0.1$$

and we reach the same conclusion.

Actually, the method used to determine the median will rarely change the conclusion. If you get something borderline, then you might want to change the method used to determine the median to verify it; otherwise, it is not likely to change anything.

Example 2:
Use the relative difference of the middles to determine whether the C++ midterm data are significantly skewed. If so, from which direction are they skewed?

Answer:
 In Example 4, §7.1, page 374, we determined the mean of the data is $\bar{x} = 837/11 \approx 76.09$.
 By the traditional sample method,

$$\#(M) = \frac{22}{2} = 11 \text{ and } M = \frac{77 + 79}{2} = 78.$$

From this,
$$\Delta_m = \frac{837/11 - 78}{78} = -\frac{7}{286} \approx -0.024476.$$
And we conclude the data are not significantly skewed.

By the restricted sample method, as both $x_{\min} = 46$ and $x_{\max} = 95$ are unique,
$$\#(M) = \frac{1+25}{2} = 13.$$
Now, $x_{(13)} = 79$ and $\#(79) = (12+13)/2 = 12.5 < 13$. We conclude $y_1 = 79$, $m_1 = 12.5$, $y_2 = 80$ and $m_2 = (14+15)/2 = 14.5$. By interpolation,
$$M = \frac{79 \cdot (14.5 - 13) + 80 \cdot (13 - 12.5)}{14.5 - 12.5} = \frac{118.5 + 40}{2} = \frac{317}{4} = 79.25.$$
Whereby we get
$$\Delta_m = \frac{837/11 - 317/4}{317/4} = -\frac{139}{3487} \approx -0.039862.$$

And we again conclude there is no significant skewing in the data.

Example 3:
Determine whether the call center data are significantly skewed. If so, from which direction are they skewed?

The tally table for the data is

24: 4	31: 1	37: 2	43: 2	48: 2	71: 1	84: 1
26: 2	32: 1	38: 2	44: 1	49: 1	72: 1	88: 1
27: 2	33: 1	39: 1	45: 2	50: 1	75: 1	91: 1
29: 3	35: 2	40: 1	46: 1	59: 1	76: 1	112: 1
30: 2	36: 1	42: 2	47: 1	66: 2	79: 1	

This makes the mean
$$\bar{x} = \frac{24 \cdot 4 + \cdots + 112 \cdot 1}{50} = \frac{2311}{50} = 46.22.$$

7.2. RELATIVE DIFFERENCE OF THE MIDDLES

By the traditional sample method

$$\#(M) = \frac{50}{2} = 25 \text{ and } M = \frac{40+42}{2} = 41.$$

This means

$$\Delta_m = \frac{46.22 - 41}{41} = \frac{5.22}{41} = \frac{261}{2050} \approx 0.127317 > 0.1$$

which means the data are skewed from the right.

The restricted sample method begins with determining the effective addresses for the sample extremes.

$$\#(x_{\min}) = \frac{1+4}{2} = 2.5 \text{ and } \#(x_{\max}) = 50.$$

This makes the effective address of the median

$$\#(M) = \frac{2.5 + 50}{2} = \frac{105}{4} = 26.25.$$

Now, $x_{(26)} = 42$ and $\#(42) = (26+27)/2 = 26.5 > 26.25$. Thereby, $y_2 = 42$, $m_2 = 26.5$, $y_1 = 40$ and $m_1 = 25$. By interpolation,

$$M = \frac{40 \cdot (26.5 - 26.25) + 42 \cdot (26.25 - 25)}{26.5 - 25}$$
$$= \frac{10 + 52.5}{1.5} = \frac{125}{3} \approx 41.6667.$$

We conclude

$$\Delta_m = \frac{46.22 - 125/3}{125/3} = \frac{683}{6250} = 0.10928 > 0.1$$

which means the data are skewed from the right.

The skewing is mild in the last example, but is still significant.

Exercises 7.2

1. One day while walking through a field you notice there seem to be a lot of stones lying on the ground. Curious about what the average size of a stone is in the field you randomly collect 30 stones. You weigh each stone and record its weight in grams. The following table shows the raw data.

110	126	75	110	126	94
90	112	111	139	131	92
102	103	96	75	67	101
104	111	100	65	92	93
106	104	93	115	109	95

   Use the relative difference of the middles to determine whether there is significant skewing. If so, give the direction of that skewing.

2. The scores, in the order graded, on a recent midterm examination in Statistics, a class with 30 students, are given below.

70	80	99	98	85	89
87	79	83	38	69	70
60	69	78	40	75	56
70	51	99	69	95	86
57	53	47	50	55	81

   Use the relative difference of the middles to determine whether there is significant skewing. If so, give the direction of that skewing.

3. Below is a sample of the monthly salary for 30 randomly selected people in mid-management in thousands of dollars.

3.08	3.18	3.13	7.41	3.73	3.05
4.17	3.58	3.36	3.27	3.36	4.74
3.32	3.61	4.26	3.02	3.45	4.06
6.73	4.72	3.13	3.15	3.70	3.59
3.12	3.61	5.03	3.53	3.20	3.32

7.2. RELATIVE DIFFERENCE OF THE MIDDLES

Use the relative difference of the middles to determine whether there is significant skewing. If so, give the direction of that skewing.

4. A pair of dice, one red and one green, is rolled in the following manner. On the first toss both dice are tossed. Then the red die is tossed while the green die keeps its value to get the second sum. Then the green die is tossed while the red die keeps its value. You continue tossing alternately the red then the green, die and generate the following table.

8	11	8	6	10	9	9
7	2	5	10	11	8	5
3	6	9	7	7	9	10
7	3	3	7	9	8	7
5	4	8	11	11	10	10

Use the relative difference of the middles to determine whether there is significant skewing. If so, give the direction of that skewing.

5. The following random sample of 72 observations comes from a gamma distribution, so $\mu_{min} = 0$ and μ_{max} has no upper limit. The raw data are

2.15	1.92	1.92	2.20	3.67	4.55	3.32	2.71	4.87
2.09	1.13	1.09	2.12	4.89	3.03	3.17	4.74	4.44
4.61	2.78	2.97	2.78	2.88	3.66	2.55	4.40	3.54
2.18	3.43	3.84	4.60	1.76	2.85	2.27	1.38	1.24
4.22	1.67	4.51	3.50	6.69	5.92	6.42	0.95	4.57
3.10	3.26	2.95	2.45	1.95	2.06	3.10	2.03	4.98
0.86	2.61	1.92	2.78	1.01	2.64	3.14	0.86	3.68
2.71	3.06	4.33	4.63	2.47	2.92	1.38	2.81	5.71

Use the relative difference of the middles to determine whether there is significant skewing. If so, give the direction of that skewing.

7.3 The Method of Differences

THIS SECTION, AND the next, concern statistics based upon quartiles. Thus, there are three methods which can be used to create them: median-of-medians, traditional sample and restricted proper. In this section we consider the method of differences.

The **method of differences** uses four statistics based upon the five-number summary. The five-number summary are the statistics

$$x_{\min} \quad Q_1 \quad M \quad Q_3 \quad x_{\max}.$$

We could have discussed it along with box-and-whiskers diagrams but decided to wait as the statistics are used for determining the skewing of the data.

The **outer differences** are the lengths between each extreme and its closest quartile. The **inner differences** are the lengths between the median and each quartile. Those differences which lie to the left of the median on the number line are **left differences**. We use $\boldsymbol{L_O}$ and $\boldsymbol{L_I}$ to denote the outer and inner left difference, respectively. Those differences which lie to the right of the median on the number line are **right differences**. We use $\boldsymbol{R_I}$ and $\boldsymbol{R_O}$ to denote the inner and outer right difference, respectively. Thus, the five-number summary yields the following four consecutive differences:

$$\underbrace{x_{\min}\ Q_1}_{L_O} \quad \underbrace{Q_1\ M}_{L_I} \quad \underbrace{M\ Q_3}_{R_I} \quad \underbrace{Q_3\ x_{\max}}_{R_O},$$

where we always do the subtraction right minus left. The inner differences are corresponding differences, as are the outer differences. We compare corresponding differences.

7.3. THE METHOD OF DIFFERENCES

1. If the smaller difference is 90% or more of the larger difference then the corresponding differences are considered equal.

2. The equality of one pair of corresponding differences cannot be used to generate inconsistent differences even if they are only equal by the 90% rule.

3. If both left differences are consistently larger than their corresponding right differences then the data are skewed from the left.

4. If both right differences are consistently larger than their corresponding left differences then the data are skewed from the right.

5. If the differences are inconsistent but the largest difference is at least 90% of the sum of the differences on the other side then the data are skewed from the side having this largest difference.

6. In all other cases, the data are considered to be balanced.

Example 1:
Use the method of differences to determine whether the dice roll data are skewed or balanced. If they are skewed then from which direction?

Answer:
The frequency table for the data is

2: 1   4: 4   6: 7   8: 7   10: 5   12: 1
3: 5   5: 6   7: 8   9: 5   11: 1

By the median-of-medians method, is

$$\#(M) = \frac{1+50}{2} = 25.5 \text{ and } M = \frac{x_{(25)} + x_{(26)}}{2} = \frac{7+7}{2} = 7.$$

The highest integer k such that $x_{(k)} \leq M$ is $k = 31$, so
$$\#(Q_1) = \frac{1+31}{2} = 16 \text{ and } Q_1 = x_{(16)} = 5.$$
The smallest integer j such that $x_{(j)} \geq M$ is $j = 24$ so
$$\#(Q_3) = \frac{24+50}{2} = 37 \text{ and } Q_3 = x_{(37)} = 8.$$
This gives the following 5-number summary
$$2; \quad 5; \quad 7; \quad 8; \quad 12.$$
The above makes
$$L_O = 3; \quad L_I = 2; \quad R_I = 1; \quad R_O = 4.$$
We observe $L_O < R_O$ but $L_I > R_I$. The corresponding differences are inconsistent. The largest difference is $R_O = 4 < 0.9 \cdot (3+2) = 4.5$. We conclude that the data must be considered balanced.

The traditional sample method agrees that $M = 7$ because $50/2 = 25$ and both $x_{(25)}$ and $x_{(26)}$ equal 7. The quartiles are computed as
$$\#(Q_1) = \frac{50}{4} = 12.5 \text{ and } Q_1 = x_{(13)} = 5$$
and
$$\#(Q_3) = 3\#(Q_1) = 37.5 \text{ and } Q_3 = x_{(38)} = 8.$$
We happen to get the same values for the 5-number summary, so the conclusion must be the same.

The restricted proper method considers the effective addresses of the sample extremes. $\#(x_{\min}) = 1$ and $\#(x_{\max}) = 50$ because 2 and 12 are both unique. Whence,
$$\#(Q_1) = \frac{3+50}{4} = 13.25$$
$$\#(M) = \frac{1+50}{2} = 25.5$$
$$\#(Q_3) = \frac{1+150}{4} = 37.75.$$

7.3. THE METHOD OF DIFFERENCES

For calculating Q_1 we observe that $x_{(13)} = 5$ and $\#(5) = (11+16)/2 = 13.5 > 13.25$. Therefore, $y_2 = 5$, $m_2 = 13.5$, $y_1 = 4$ and $m_1 = (7+10)/2 = 8.5$. By interpolation,

$$Q_1 = \frac{4 \cdot (13.5 - 13.25) + 5 \cdot (13.25 - 8.5)}{13.5 - 8.5} = \frac{1 + 23.75}{5} = \frac{99}{20} = 4.95.$$

In calculating M we see that $x_{(25)} = 7$ and $\#(7) = (24+31)/2 = 27.5 > 25.5$. Whereby, $y_2 = 7$, $m_2 = 27.5$, $y_1 = 6$ and $m_1 = (17+23)/2 = 20$. By interpolation,

$$M = \frac{6 \cdot (27.5 - 25.5) + 7 \cdot (25.5 - 20)}{27.5 - 20} = \frac{12 + 38.5}{7.5} = \frac{101}{15} \approx 6.733333.$$

To calculate Q_3 we note $x_{(37)} = 8$ and $\#(8) = (32+38)/2 = 35 < 37.75$. Thereby, $y_1 = 8$, $m_1 = 35$, $y_2 = 9$ and $m_2 = (39+43)/2 = 41$. Interpolation gives

$$Q_3 = \frac{8 \cdot (41 - 37.75) + 9 \cdot (37.75 - 35)}{41 - 35}$$
$$= \frac{26 + 24.75}{6} = \frac{203}{24} \approx 8.458333.$$

The 5-number summary is

$$2; \quad \frac{99}{20}; \quad \frac{101}{15}; \quad \frac{203}{24}; \quad 12.$$

This makes the following differences

$$L_O = \frac{59}{20} = 2.95 \qquad R_O = \frac{85}{24} \approx 3.541667$$
$$L_I = \frac{107}{60} \approx 1.783333 \qquad R_I = \frac{69}{40} = 1.725.$$

This makes $L_O < R_O$ and $L_I \approx R_I$. We conclude that there is mild skewing from the right as the approximate equality cannot be used to create a inconsistency.

The above example shows that the way the quartiles are calculated can have an effect on the final conclusion. That is the problem with this statistic.

Example 2:
Determine if the C++ midterm data is skewed or balanced using the method of differences.

Answer:
The sorted data are

$$\begin{array}{cccccccccc} 46 & 55 & 62 & 67 & 68 & 68 & 72 & 74 & 75 & 76 & 77 \\ 79 & 79 & 80 & 80 & 81 & 82 & 86 & 88 & 91 & 94 & 95 \end{array}$$

From the median-of-medians method, $\#(M) = (1+22)/2 = 11.5$, so $M = (77+79)/2 = 78$. We conclude $j = 11$ and $k = 12$. Therefore,

$$\#(Q_1) = \frac{1+11}{2} = 6 \text{ and } \#(Q_3) = \frac{12+22}{2} = 17.$$

This means $Q_1 = 68$ and $Q_3 = 82$. The five-number summary is

$$46; \quad 68; \quad 78; \quad 82; \quad 95.$$

We conclude that the differences are

$$L_O = 22; \quad L_I = 10; \quad R_I = 4; \quad R_O = 13.$$

Clearly, $0.9 \cdot 22 = 19.8 > 13$ and $0.9 \cdot 10 = 9 > 4$ and the data are skewed from the left. In fact, $22 > 0.9 \cdot (3 + 14) = 15.3$ and the data are skewed from the left for more than just one reason.

The quartiles obtained from the traditional sample method agree with those of the median-of-medians method in this case. (Check this!) We conclude the same result.

7.3. THE METHOD OF DIFFERENCES

Both sample extremes are unique, so

$$\#(Q_1) = \frac{3+22}{4} = 6.25;$$

$$\#(M) = \frac{1+22}{2} = 11.5;$$

$$\#(Q_3) = \frac{1+66}{4} = 16.75.$$

Now, $x_{(6)} = 68$ and $\#(6) = (5+6)/2 = 5.5 < 6.25$. We conclude that $y_1 = 68$, $m_1 = 5.5$, $y_2 = 72$ and $m_2 = 7$. Using interpolation we get

$$Q_1 = \frac{68 \cdot (7 - 6.25) + 72 \cdot (6.25 - 5.5)}{7 - 5.5} = \frac{51 + 54}{1.5} = 70.$$

Also, $x_{(11)} = 77$ and $\#(77) = 11 < 11.5$. This means that $y_1 = 77$, $m_1 = 11$, $y_2 = 79$ and $m_2 = (12+13)/2 = 12.5$. From interpolation we calculate

$$M = \frac{77 \cdot (12.5 - 11.5) + 79 \cdot (11.5 - 11)}{12.5 - 11}$$
$$= \frac{77 + 39.5}{1.5} = \frac{233}{3} \approx 77.6667.$$

Finally, $x_{(16)} = 81$ and $\#(81) = 16 < 16.75$. Whereby, $y_1 = 81$, $m_1 = 16$, $y_2 = 82$ and $m_2 = 17$. The difference in effective positions is 1, so linear interpolation gives

$$Q_3 = 81 \cdot (17 - 16.75) + 82 \cdot (16.75 - 16) = 20.25 + 61.5 = 81.75.$$

The five-number summary is

$$46; \quad 70; \quad \frac{233}{3}; \quad \frac{327}{4}; \quad 95$$

and the differences are

$$L_O = 24; \qquad R_O = \frac{53}{4} = 13.25.$$

$$L_I = \frac{23}{3} \approx 7.6667; \quad R_I = \frac{49}{12} \approx 4.0833;$$

We reach the same conclusion as before.

Exercises 7.3

1. One day while walking through a field you notice there seem to be a lot of stones lying on the ground. Curious about what the average size of a stone is in the field you randomly collect 30 stones. You weigh each stone and record its weight in grams. The following table shows the raw data.

110	126	75	110	126	94
90	112	111	139	131	92
102	103	96	75	67	101
104	111	100	65	92	93
106	104	93	115	109	95

   Use the method of differences to determine whether there is significant skewing. If so, give the direction of that skewing.

2. The scores, in the order graded, on a recent midterm examination in Statistics, a class with 30 students, are given below.

70	80	99	98	85	89
87	79	83	38	69	70
60	69	78	40	75	56
70	51	99	69	95	86
57	53	47	50	55	81

   Use the method of differences to determine whether there is significant skewing. If so, give the direction of that skewing.

3. Below is a sample of the monthly salary for 30 randomly selected people in mid-management in thousands of dollars.

3.08	3.18	3.13	7.41	3.73	3.05
4.17	3.58	3.36	3.27	3.36	4.74
3.32	3.61	4.26	3.02	3.45	4.06
6.73	4.72	3.13	3.15	3.70	3.59
3.12	3.61	5.03	3.53	3.20	3.32

Use the method of differences to determine whether there is significant skewing. If so, give the direction of that skewing.

4. A pair of dice, one red and one green, is rolled in the following manner. On the first toss both dice are tossed. Then the red die is tossed while the green die keeps its value to get the second sum. Then the green die is tossed while the red die keeps its value. You continue tossing alternately the red then the green, die and generate the following table.

8	11	8	6	10	9	9
7	2	5	10	11	8	5
3	6	9	7	7	9	10
7	3	3	7	9	8	7
5	4	8	11	11	10	10

Use the method of differences to determine whether there is significant skewing. If so, give the direction of that skewing.

5. The following random sample of 72 observations comes from a gamma distribution, so $\mu_{\text{min}} = 0$ and μ_{max} has no upper limit. The raw data are

2.15	1.92	1.92	2.20	3.67	4.55	3.32	2.71	4.87
2.09	1.13	1.09	2.12	4.89	3.03	3.17	4.74	4.44
4.61	2.78	2.97	2.78	2.88	3.66	2.55	4.40	3.54
2.18	3.43	3.84	4.60	1.76	2.85	2.27	1.38	1.24
4.22	1.67	4.51	3.50	6.69	5.92	6.42	0.95	4.57
3.10	3.26	2.95	2.45	1.95	2.06	3.10	2.03	4.98
0.86	2.61	1.92	2.78	1.01	2.64	3.14	0.86	3.68
2.71	3.06	4.33	4.63	2.47	2.92	1.38	2.81	5.71

Use the method of differences to determine whether there is significant skewing. If so, give the direction of that skewing.

7.4 Bowley's Skewness Coefficient

ONE OF THE FIRST attempts at determining a skewness coefficient was done by Bowley in the 1930s. Bowley's coefficient determines the skewness in the core, the middle half of the data. If the method of diffeences has inconsistent differences then even the sign of skewness may differ from the overall sample.

Bowley's skewness coefficient, denoted B, is defined as

$$B = \frac{R_I - L_I}{R_Q} = \frac{Q_1 + Q_3 - 2M}{Q_3 - Q_1},$$

Where R_I and L_I are the right and left inner differences, respectively, and R_Q is the interquartile range.

We now know that the standard deviation of Bowley's coefficient is finite and bounded by some inverse multiple of the square root of n, but we do not know the exact coefficient. However, there is an estimate called the **standard error of skewness**, denoted se_s, which is often used to estimate the significance level.

$$\text{se}_s = \sqrt{\frac{6}{n}}.$$

Table 7.4 shows how the standard error of skewness is used with Bowley's coefficient. As with all statistics of skewness, negative values denote skewness from the left (more high than low values) and positive values denote skewness from the right (more low than high values).

Example 1:
Determine the core skewness in the two-success data.

7.4. BOWLEY'S SKEWNESS COEFFICIENT

Table 7.4: Using se_s with Bowley's coefficient of skewness

$	B	< \text{se}_s$	not significantly skewed
$\text{se}_s \leq	B	\leq 2\text{se}_s$	mildly skewed
$	B	> 2\text{se}_s$	severely skewed

Answer:
The core skewness is measured by Bowley's skewness coefficient, so we need to determine the quartiles of the data. The frequency table is

2: 9   4: 15   6: 9   8: 4   10: 6   12: 2   15: 1   17: 1
3: 16   5: 17   7: 4   9: 5   11: 4   14: 5   16: 1   28: 1

There are 100 data, so

$$\text{se}_s = \sqrt{\frac{6}{100}} = \frac{\sqrt{6}}{10} \approx 0.244949.$$

In the median-of-medians method

$$\#(M) = \frac{1+100}{2} = 50.5 \text{ and } M = \frac{x_{(50)} + x_{(51)}}{2} = \frac{5+5}{2} = 5.$$

The largest integer j such that $x_{(j)} \leq 5$ is $j = 57$, so

$$\#(Q_1) = \frac{1+57}{2} = 29 \text{ and } Q_1 = x_{(29)} = 4.$$

The smallest integer k such that $x_{(k)} \geq 5$ is $k = 41$, so

$$\#(Q_3) = \frac{41+100}{2} = 70.5 \text{ and } Q_3 = \frac{x_{(70)} + x_{(71)}}{2} = \frac{7+8}{2} = 7.5.$$

Bowley's skewness coefficient is

$$B = \frac{4 + 7.5 - 2 \cdot 5}{7.5 - 4} = \frac{1.5}{3.5} = \frac{3}{7} \approx 0.428571.$$

Because
$$0.244949 < 0.428571 < 0.489898,$$
We conclude the interquartile data is mildly skewed from the right.

In the traditional sample method
$$\#(Q_1) = \frac{100}{4} = 25; \quad \#(M) = \frac{100}{2} = 50; \quad \#(Q_3) = 3\#(Q_1) = 75.$$
This means
$$Q_1 = \frac{x_{(25)} + x_{(26)}}{2} = \frac{3+4}{2} = 3.5$$
$$M = \frac{x_{(50)} + x_{(51)}}{2} = \frac{5+5}{2} = 5$$
$$Q_3 = \frac{x_{(75)} + x_{(76)}}{2} = \frac{9+9}{2} = 9.$$

Bowley's skewness coefficient is
$$B = \frac{3.5 + 9 - 2 \cdot 5}{9 - 3.5} = \frac{2.5}{5.5} = \frac{5}{11} \approx 0.454545.$$

Again we conclude the core is mildly skewed from the right.

In the restricted proper method,
$$\#(x_{\min}) = \frac{1+9}{2} = 5 \text{ and } \#(x_{\max}) = 100.$$
From this,
$$\#(Q_1) = \frac{115}{4} = 28.75;$$
$$\#(M) = \frac{105}{2} = 52.5;$$
$$\#(Q_3) = \frac{305}{4} = 76.25.$$

7.4. BOWLEY'S SKEWNESS COEFFICIENT

To obtain Q_1 we start with $x_{28} = 4$ and $\#(4) = (26 + 40)/2 = 33 > 28.75$. Therefore, $y_2 = 4$, $m_2 = 33$, $y_1 = 3$ and $m_1 = (10 + 25)/2 = 17.5$. Interpolating between these points gives

$$Q_1 = \frac{3 \cdot (33 - 28.75) + 4 \cdot (28.75 - 17.5)}{33 - 17.5}$$
$$= \frac{12.75 + 45}{15.5} = \frac{231}{62} \approx 3.725806.$$

For M, $x_{(52)} = 5$ and $\#(5) = (41 + 57)/2 = 49 < 52.5$. Heretofore, $y_1 = 5$, $m_1 = 49$, $y_2 = 6$ and $m_2 = (58 + 66)/2 = 62$. Through liner interpolation we calculate

$$M = \frac{5 \cdot (62 - 52.5) + 6 \cdot (52.5 - 49)}{62 - 49} = \frac{47.5 + 21}{13} = \frac{137}{26} \approx 5.269231.$$

With Q_3 we begin by noting $x_{(76)} = 9$ and $\#(9) = (75 + 79)/2 = 77 > 76.25$. Wherefore, $y_2 = 9$, $m_2 = 77$, $y_1 = 8$ and $m_1 = (71 + 74)/2 = 72.5$. By interpolation,

$$Q_3 = \frac{8 \cdot (77 - 76.25) + 9 \cdot (76.25 - 72.5)}{77 - 72.5}$$
$$= \frac{6 + 33.75}{4.5} = \frac{53}{6} \approx 8.833333.$$

Bowley's skewness coefficient is

$$B = \frac{\frac{231}{62} + \frac{53}{6} - 2 \cdot \frac{137}{26}}{\frac{53}{6} - \frac{231}{62}} = \frac{2443/1209}{475/93} = \frac{2443}{6175} \approx 0.395628.$$

Which gives the same result as before.

If in using Bowley's coefficient you get a result which is close to one of the test boundaries then you might want to obtain quartiles from another method to verify the answer. In general, there is not much difference between the methods when the value is not close to a test boundary.

Example 2:
Use Bowley's skewness coefficient to determine whether the central 50% of the C++ midterm data are balanced.

Answer:
From Example 2, §7.3, starting on page 388, we get the following quartiles

Method	Q_1	M	Q_3
Median-of-Medians	68	78	82
Traditional Sample	68	78	82
Restricted Proper	70	77.6667	81.75

For the median-of-medians and traditional sample method, Bowley's skewness coefficient is

$$B = \frac{68 + 82 - 2 \cdot 78}{82 - 68} = \frac{-6}{14} = -\frac{3}{7} \approx -0.4286.$$

For the restricted proper method

$$B = \frac{70 + 81.75 - 2 \cdot \frac{233}{3}}{81.75 - 70} = \frac{-49/12}{47/4} = -\frac{49}{141} \approx -0.3475$$

The standard error of skewness is

$$\mathrm{se}_s = \sqrt{\frac{6}{22}} = \frac{\sqrt{33}}{11} \approx 0.5222.$$

In both cases, $|B| < \mathrm{se}_s$, so the skewing is not significant.

7.4. BOWLEY'S SKEWNESS COEFFICIENT 397

Exercises 7.4

1. One day while walking through a field you notice there seem to be a lot of stones lying on the ground. Curious about what the average size of a stone is in the field you randomly collect 30 stones. You weigh each stone and record its weight in grams. The following table shows the raw data.

110	126	75	110	126	94
90	112	111	139	131	92
102	103	96	75	67	101
104	111	100	65	92	93
106	104	93	115	109	95

   Determine whether the core of the data is significantly skewed and if so, the direction of that skewing.

2. The scores, in the order graded, on a recent midterm examination in Statistics, a class with 30 students, are given below.

70	80	99	98	85	89
87	79	83	38	69	70
60	69	78	40	75	56
70	51	99	69	95	86
57	53	47	50	55	81

   Determine whether the central 50% of the data is significantly skewed and if so, the direction of that skewing.

3. Below is a sample of the monthly salary for 30 randomly selected people in mid-management in thousands of dollars.

3.08	3.18	3.13	7.41	3.73	3.05
4.17	3.58	3.36	3.27	3.36	4.74
3.32	3.61	4.26	3.02	3.45	4.06
6.73	4.72	3.13	3.15	3.70	3.59
3.12	3.61	5.03	3.53	3.20	3.32

Determine whether the interquartile data are significantly skewed and if so, the direction of that skewing.

4. A pair of dice, one red and one green, is rolled in the following manner. On the first toss both dice are tossed. Then the red die is tossed while the green die keeps its value to get the second sum. Then the green die is tossed while the red die keeps its value. You continue tossing alternately the red then the green, die and generate the following table.

8	11	8	6	10	9	9
7	2	5	10	11	8	5
3	6	9	7	7	9	10
7	3	3	7	9	8	7
5	4	8	11	11	10	10

Determine whether the core of the data is significantly skewed and if so, the direction of that skewing.

5. The following random sample of 72 observations comes from a gamma distribution, so $\mu_{min} = 0$ and μ_{max} has no upper limit. The raw data are

2.15	1.92	1.92	2.20	3.67	4.55	3.32	2.71	4.87
2.09	1.13	1.09	2.12	4.89	3.03	3.17	4.74	4.44
4.61	2.78	2.97	2.78	2.88	3.66	2.55	4.40	3.54
2.18	3.43	3.84	4.60	1.76	2.85	2.27	1.38	1.24
4.22	1.67	4.51	3.50	6.69	5.92	6.42	0.95	4.57
3.10	3.26	2.95	2.45	1.95	2.06	3.10	2.03	4.98
0.86	2.61	1.92	2.78	1.01	2.64	3.14	0.86	3.68
2.71	3.06	4.33	4.63	2.47	2.92	1.38	2.81	5.71

Determine whether the interquartile data are significantly skewed and if so, the direction of that skewing.

7.5 Pearson's Skewness Coefficient

THERE IS VERY little known about the significant levels of this next skewness coefficient. **Pearson's skewness coefficient**, denoted P, is quite similar to the difference of the middles, except for a scaling factor. By definition,

$$P = \frac{3 \cdot (\bar{x} - M)}{s},$$

where $\bar{x}$ is the mean, M is the median and s is the standard deviation. Some people use s_d, the standard deviance, here instead of the standard deviation. We will use P_d for that variation. Thus,

$$P_d = \frac{3 \cdot (\bar{x} - M)}{s_d}.$$

Example 1:
Compare both variations for Pearson's skewness coefficient to decide whether the two-success data are skewed. If so, then identify the direction of skewing.

The tally table for this data is

2: 9   4: 15   6: 9   8: 4   10: 6   12: 2   15: 1   17: 1
3: 16   5: 17   7: 4   9: 5   11: 4   14: 5   16: 1   28: 1

From this,

$$\sum_{i=1}^{100} x_i = 644 \text{ and } \sum_{i=1}^{100} x_i^2 = 5932.$$

Therefore,

$$\bar{x} = 6.44; \quad V = \frac{100 \cdot 5932 - 644^2}{100 \cdot 99} = \frac{1352}{75}; \quad s = \frac{26\sqrt{6}}{15} \approx 4.245782.$$

By the traditional sample method,

$$\#(M) = \frac{100}{2} \text{ and } M = \frac{x_{(50)} + x_{(51)}}{2} = \frac{5+5}{2} = 5.$$

This gives the following absolute difference tallies

0: 17	2: 20	4: 5	6: 4	9: 5	11: 1	23: 1
1: 24	3: 13	5: 6	7: 2	10: 1	12: 1	

Using this,

$$\sum_{i=1}^{100} d_i^2 = 1992 \text{ and } \mathcal{V} = \frac{1992}{99} = \frac{664}{33} \text{ so } s_d = \frac{2\sqrt{5478}}{33} \approx 4.485667.$$

This means

$$P = \frac{3 \cdot (6.44 - 5)}{4.245782} \approx 1.0175 \text{ and } P_d = \frac{3 \cdot (6.44 - 5)}{4.485667} \approx 0.96307$$

The data are positively skewed; that is, skewed from the right.

By the restricted proper method,

$$\#(x_{\min}) = (1+9)/2 = 5 \text{ and } \#(x_{\max}) = 100,$$

so,

$$\#(M) = \frac{5 + 100}{2} = 52.5.$$

We note that $x_{(52)} = 5$ and $\#(5) = (41+57)/2 = 49 < 52.5$. Hence, $y_1 = 5$, $m_1 = 49$, $y_2 = 6$ and $m_2 = (58+66)/2 = 62$. From linear interpolation we calculate

$$M = \frac{5 \cdot (62 - 52.5) + 6 \cdot (52.5 - 49)}{62 - 49} = \frac{47.5 + 21}{13} = \frac{137}{26} \approx 5.26923.$$

This gives the following approximate absolute difference table:

0.26923: 17	2.26923: 16	4.73077: 6	9.73077: 1
0.73077: 9	2.73077: 4	5.73077: 4	10.73077: 1
1.26923: 15	3.26923: 9	6.73077: 2	11.73077: 1
1.73077: 4	3.73077: 5	8.73077: 5	22.73077: 1

7.5. PEARSON'S SKEWNESS COEFFICIENT

With the above,

$$\sum_{i=1}^{100} d_i^2 \approx 1921.71024; \quad \mathcal{V} \approx \frac{1921.71024}{99} \approx 19.41121; \text{ so, } s_d \approx 4.40582.$$

This makes

$$P = \frac{3 \cdot (6.44 - 5.26923)}{4.245782} \approx 0.82725$$

and

$$P_d = \frac{3 \cdot (6.44 - 5.26923)}{4.40582} \approx 0.79720.$$

We conclude the data are positively skewed; that is, they are skewed from the right.

Example 2:
Determine whether the C++ midterm data are skewed using both variations of Pearson's skewness coefficient. If so, from which direction is the skewing?

Answer:
The sorted data are

| 46 | 55 | 62 | 67 | 68 | 68 | 72 | 74 | 75 | 76 | 77 |
| 79 | 79 | 80 | 80 | 81 | 82 | 86 | 88 | 91 | 94 | 95 |

From which we see

$$\sum_{i=1}^{22} x_i = 1675; \quad \sum_{i=1}^{22} x_i^2 = 130561.$$

There are 22 scores, so

$$\bar{x} = \frac{1675}{22}; \mathcal{V} = \frac{22 \cdot 130561 - 1675^2}{22 \cdot 21} = \frac{3177}{22} \text{ and } s \approx 12.017033.$$

By the traditional sample method

$$\#(M) = \frac{22}{2} = 11 \text{ and } M = \frac{x_{(11)} + x_{(12)}}{2} = \frac{77 + 79}{2} = 78.$$

This gives the following table of absolute differences

1: 3	3: 2	6: 1	10: 3	13: 1	17: 1	32: 1
2: 3	4: 2	8: 1	11: 1	16: 2	23: 1	

From the above

$$\sum_{i=1}^{22} d_i^2 = 3109; \quad \mathcal{V} = \frac{3109}{21} \text{ and } s_{\mathrm{d}} \approx 12.167482.$$

We conclude,

$$P = \frac{3 \cdot (76.13636 - 78)}{12.017033} \approx -0.46525$$

and

$$P_{\mathrm{d}} = \frac{3 \cdot (76.13636 - 78)}{12.167482} \approx -0.45950.$$

This indicates that the data are skewed from the left; that is, there are more higher values than lower values.

Using the restricted proper method we note that $x_{\min} = 46$ and $x_{\max} = 95$ are unique, so the effective address of M is the same as in the traditional sample method. Now, $x_{(11)} = 77$ and $\#(77) = 11 < 11.5$. Whence, $y_1 = 77$, $m_1 = 11$, $y_2 = 79$ and $m_2 = (12 + 13)/2 = 12.5$. Linear interpolation gives

$$M = \frac{77 \cdot (12.5 - 11.5) + 79 \cdot (11.5 - 11)}{12.5 - 11}$$
$$= \frac{77 + 39.5}{1.5} = \frac{233}{3} = 77.666667.$$

The absolute differences are

2/3	4/3	4/3	5/3	7/3	7/3	8/3	10/3	11/3	13/3	17/3
25/3	29/3	29/3	31/3	32/3	40/3	47/3	49/3	52/3	68/3	95/3

7.5. PEARSON'S SKEWNESS COEFFICIENT

Therefrom,

$$\sum_{i=1}^{22} d_i^2 = \frac{27757}{9}; \mathcal{V} = \frac{27757}{189} \text{ and } s_\mathrm{d} \approx 12.118681$$

Therefore,

$$P = \frac{3 \cdot (1675/22 - 233/3)}{12.017033} \approx {-0.38203}$$

and

$$P_\mathrm{d} = \frac{3 \cdot (1675/22 - 233/3)}{12.118681} \approx {-0.37883}.$$

The data are negatively skewed; that is, skewed from the left.

Exercises 7.5

1. One day while walking through a field you notice there seem to be a lot of stones lying on the ground. Curious about what the average size of a stone is in the field you randomly collect 30 stones. You weigh each stone and record its weight in grams. The following table shows the raw data.

110	126	75	110	126	94
90	112	111	139	131	92
102	103	96	75	67	101
104	111	100	65	92	93
106	104	93	115	109	95

   Based upon both variations of Pearson's skewness coefficient, from which direction are the data skewed. Do not worry about the significance level.

2. The scores, in the order graded, on a recent midterm examination in Statistics, a class with 30 students, are given below.

70	80	99	98	85	89
87	79	83	38	69	70
60	69	78	40	75	56
70	51	99	69	95	86
57	53	47	50	55	81

Based upon both variations of Pearson's skewness coefficient, from which direction are the data skewed. Do not worry about the significance level.

3. Below is a sample of the monthly salary for 30 randomly selected people in mid-management in thousands of dollars.

3.08	3.18	3.13	7.41	3.73	3.05
4.17	3.58	3.36	3.27	3.36	4.74
3.32	3.61	4.26	3.02	3.45	4.06
6.73	4.72	3.13	3.15	3.70	3.59
3.12	3.61	5.03	3.53	3.20	3.32

Based upon both variations of Pearson's skewness coefficient, from which direction are the data skewed. Do not worry about the significance level.

4. A pair of dice, one red and one green, is rolled in the following manner. On the first toss both dice are tossed. Then the red die is tossed while the green die keeps its value to get the second sum. Then the green die is tossed while the red die keeps its value. You continue tossing alternately the red then the green, die and generate the following table.

8	11	8	6	10	9	9
7	2	5	10	11	8	5
3	6	9	7	7	9	10
7	3	3	7	9	8	7
5	4	8	11	11	10	10

7.6. FISHER'S SKEWNESS COEFFICIENT

Based upon both variations of Pearson's skewness coefficient, from which direction are the data skewed. Do not worry about the significance level.

5. The following random sample of 72 observations comes from a gamma distribution, so $\mu_{\min} = 0$ and $\mu_{\max}$ has no upper limit. The raw data are

2.15	1.92	1.92	2.20	3.67	4.55	3.32	2.71	4.87
2.09	1.13	1.09	2.12	4.89	3.03	3.17	4.74	4.44
4.61	2.78	2.97	2.78	2.88	3.66	2.55	4.40	3.54
2.18	3.43	3.84	4.60	1.76	2.85	2.27	1.38	1.24
4.22	1.67	4.51	3.50	6.69	5.92	6.42	0.95	4.57
3.10	3.26	2.95	2.45	1.95	2.06	3.10	2.03	4.98
0.86	2.61	1.92	2.78	1.01	2.64	3.14	0.86	3.68
2.71	3.06	4.33	4.63	2.47	2.92	1.38	2.81	5.71

Based upon both variations of Pearson's skewness coefficient, from which direction are the data skewed. Do not worry about the significance level.

7.6 Fisher's Skewness Coefficient

About the same time as Bowley's skewness coefficient was being developed, Bowley's PhD student, Fisher was developing another skewness coefficient. **Fisher's skewness coefficient**, denoted γ, is the most commonly used skewness coefficient. All of the skewness coefficients have been extremely difficult to analyze. In fact, it wasn't until 1987 that it was proven that the variance of Fisher's skewness coefficient is

$$V(\gamma) = \text{se}_s^2 = \frac{6}{n},$$

so the standard error of skewness is the standard deviation of Fisher's skewness coefficient. This means that it is possible to calculate true significance levels for this statistic. We will, however, just use the same test that we used for Bowley's skewness coefficient, at least for now.

The formula for γ is

$$\gamma = \frac{\sqrt{n} \sum_{i=1}^{n}(x-\mu)^3}{\left(\sum_{i=1}^{n}(x-\mu)^2\right)^{3/2}}.$$

For a sample, we use $\overline{x}$ in place of μ.

Recall:

$$\sum_{i=1}^{n}(x-\overline{x})^2 = \frac{n\sum_{i=1}^{n}x_i^2 - \left(\sum_{i=1}^{n}x_i\right)^2}{n} \text{ and } \overline{x} = \frac{\sum_{i=1}^{n}x_i}{n}.$$

We will use these in deriving a calculating formula for γ.

$$\sum_{i=1}^{n}(x_i-\overline{x})^3 = \sum_{i=1}^{n}(x_i^3 - 3x_i^2\overline{x} + 3x_i\overline{x}^2 - \overline{x}^3)$$

$$= \sum_{i=1}^{n}x_i^3 - 3\overline{x}\sum_{i=1}^{n}x_i^2 + 3\overline{x}^2\sum_{i=1}^{n}x_i - \overline{x}^3\sum_{i=1}^{n}1$$

$$= \sum_{i=1}^{n}x_i^3 - 3\left(\sum_{i=1}^{n}x_i\right)\left(\sum_{i=1}^{n}x_i^2\right)/n$$

$$+ 3\left(\sum_{i=1}^{n}x_i\right)^3/n^2 - n\left(\sum_{i=1}^{n}x_i\right)^3/n^3$$

7.6. FISHER'S SKEWNESS COEFFICIENT

$$= \sum_{i=1}^{n} x_i^3 - \frac{3}{n}\left(\sum_{i=1}^{n} x_i\right)\left(\sum_{i=1}^{n} x_i^2\right) + \frac{2}{n^2}\left(\sum_{i=1}^{n} x_i\right)^3$$

$$= \frac{n^2 \sum_{i=1}^{n} x_i^3 - 3n\left(\sum_{i=1}^{n} x_i\right)\left(\sum_{i=1}^{n} x_i^2\right) + 2\left(\sum_{i=1}^{n} x_i\right)^3}{n^2}.$$

From the above we get

$$\gamma = \frac{\sqrt{n}\sum_{i=1}^{n}(x-\mu)^3}{\left(\sum_{i=1}^{n}(x-\mu)^2\right)^{3/2}}$$

$$= \frac{\sqrt{n}\left(\dfrac{n^2 \sum_{i=1}^{n} x_i^3 - 3n\left(\sum_{i=1}^{n} x_i\right)\left(\sum_{i=1}^{n} x_i^2\right) + 2\left(\sum_{i=1}^{n} x_i\right)^3}{n^2}\right)}{\left(\dfrac{n\sum_{i=1}^{n} x_i^2 - \left(\sum_{i=1}^{n} x_i\right)^2}{n}\right)^{3/2}}$$

Which simplifies to

$$\gamma = \frac{n^2 \sum_{i=1}^{n} x_i^3 - 3n\left(\sum_{i=1}^{n} x_i\right)\left(\sum_{i=1}^{n} x_i^2\right) + 2\left(\sum_{i=1}^{n} x_i\right)^3}{\left(n\sum_{i=1}^{n} x_i^2 - \left(\sum_{i=1}^{n} x_i\right)^2\right)^{3/2}}$$

Example 1:
Use Fisher's coefficient of skewness to determine whether the two-success data are significantly skewed and if so, from which direction.

Answer:
The tally table for this data is

y_i	f_i	$y_i \cdot f_i$	$y_i^2 \cdot f_i$	$y_i^3 \cdot f_i$	y_i	f_i	$y_i \cdot f_i$	$y_i^2 \cdot f_i$	$y_i^3 \cdot f_i$
2	9	18	36	72	10	6	60	600	6000
3	16	48	144	432	11	4	44	484	5324
4	15	60	240	960	12	2	24	288	3456
5	17	85	425	2125	14	5	70	980	13720
6	9	54	324	1944	15	1	15	225	3375
7	4	28	196	1372	16	1	16	256	4096
8	4	32	256	2048	17	1	17	289	4913
9	5	45	405	3645	28	1	28	784	21952

$$\sum f = 100 \quad \sum yf = 644 \quad \sum y^2 f = 5932 \quad \sum y^3 f = 75434$$

The numerator is

$$100^2 \cdot 75434 - 3 \cdot 100 \cdot 644 \cdot 5932 + 2 \cdot 644^3 = 142457568.$$

Inside the parentheses we have

$$100 \cdot 5932 - 644^2 = 178464$$

so the denominator is

$$178464^{3/2} = 178464 \cdot \sqrt{178464} = 9280128 \cdot \sqrt{66}.$$

Combining

$$\gamma = \frac{142457568}{9280128 \cdot \sqrt{66}} = \frac{134903 \cdot \sqrt{66}}{580008} \approx 1.889555.$$

The standard error of skewness is

$$\text{se}_s = \sqrt{\frac{6}{100}} = \frac{\sqrt{6}}{10} \approx 0.244949.$$

7.6. FISHER'S SKEWNESS COEFFICIENT

This makes
$$\frac{\gamma}{\text{se}_s} = \frac{674515}{290004} \cdot \sqrt{11} \approx 7.714077.$$

We conclude there is a significant positive skewing; that is, there are considerably more smaller values than larger values so the data are skewed from the right.

Example 2:

Use Fisher's coefficient of skewness to determine whether the C++ midterm data are significantly skewed and if so, from which direction.

Answer:

The tally table for this data is

y_i	f_i	$y_i \cdot f_i$	$y_i^2 \cdot f_i$	$y_i^3 \cdot f_i$	y_i	f_i	$y_i \cdot f_i$	$y_i^2 \cdot f_i$	$y_i^3 \cdot f_i$
46	1	46	2116	97336	79	2	158	12482	986078
55	1	55	3025	166375	80	2	160	12800	1024000
62	1	62	3844	238328	81	1	81	6561	531441
67	1	67	4489	300763	82	1	82	6724	551368
68	2	136	9248	728864	86	1	86	7396	636056
72	1	72	5184	373248	88	1	88	7744	681472
74	1	74	5476	405224	91	1	91	8281	753571
75	1	75	5625	421875	94	1	94	8836	830584
76	1	76	5776	438976	95	1	95	9025	857375
77	1	77	5929	456533					

$$\sum f = 22 \quad \sum yf = 1675 \quad \sum y^2 f = 130561 \quad \sum y^3 f = 10379467$$

The numerator is
$$22^2 \cdot 10379467 - 3 \cdot 22 \cdot 1675 \cdot 130561 + 2 \cdot 1675^3 = {}^-11012772.$$

In side the parentheses we get
$$22 \cdot 130561 - 1675^2 = 66717$$

so the denominator is
$$66717^{3/2} = 66717 \cdot \sqrt{66717} = 200151 \cdot \sqrt{7413}.$$

This makes
$$\gamma = \frac{-11012772}{200151 \cdot \sqrt{7413}} = -\frac{3670924 \cdot \sqrt{7413}}{494573121} \approx -0.639060.$$

The standard error of skewness is
$$\text{se}_s = \sqrt{\frac{6}{22}} = \frac{\sqrt{33}}{11} \approx 0.522233.$$

Which means that
$$\frac{\gamma}{\text{se}_s} = -\frac{3670924 \cdot \sqrt{27181}}{494573121} \approx -1.223707.$$

We conclude that the negative skewing of the data is borderline; it is slightly skewed from the left so there are more higher values than lower values.

Exercises 7.6

1. One day while walking through a field you notice there seem to be a lot of stones lying on the ground. Curious about what the average size of a stone is in the field you randomly collect 30 stones. You weigh each stone and record its weight in grams. The following table shows the raw data.

110	126	75	110	126	94
90	112	111	139	131	92
102	103	96	75	67	101
104	111	100	65	92	93
106	104	93	115	109	95

   Based upon Fisher's skewness coefficient, from which direction are the data skewed. Use the Bowley rule-of-thumb to determine the significance level.

7.6. FISHER'S SKEWNESS COEFFICIENT

2. The scores, in the order graded, on a recent midterm examination in Statistics, a class with 30 students, are given below.

70	80	99	98	85	89
87	79	83	38	69	70
60	69	78	40	75	56
70	51	99	69	95	86
57	53	47	50	55	81

   Based upon Fisher's skewness coefficient, from which direction are the data skewed. Use the Bowley rule-of-thumb to determine the significance level.

3. Below is a sample of the monthly salary for 30 randomly selected people in mid-management in thousands of dollars.

3.08	3.18	3.13	7.41	3.73	3.05
4.17	3.58	3.36	3.27	3.36	4.74
3.32	3.61	4.26	3.02	3.45	4.06
6.73	4.72	3.13	3.15	3.70	3.59
3.12	3.61	5.03	3.53	3.20	3.32

   Based upon Fisher's skewness coefficient, from which direction are the data skewed. Use the Bowley rule-of-thumb to determine the significance level.

4. A pair of dice, one red and one green, is rolled in the following manner. On the first toss both dice are tossed. Then the red die is tossed while the green die keeps its value to get the second sum. Then the green die is tossed while the red die keeps its value. You continue tossing alternately the red then the green, die and generate the following table.

8	11	8	6	10	9	9
7	2	5	10	11	8	5
3	6	9	7	7	9	10
7	3	3	7	9	8	7
5	4	8	11	11	10	10

Based upon Fisher's skewness coefficient, from which direction are the data skewed. Use the Bowley rule-of-thumb to determine the significance level.

5. The following random sample of 72 observations comes from a gamma distribution, so $\mu_{min} = 0$ and μ_{max} has no upper limit. The raw data are

2.15	1.92	1.92	2.20	3.67	4.55	3.32	2.71	4.87
2.09	1.13	1.09	2.12	4.89	3.03	3.17	4.74	4.44
4.61	2.78	2.97	2.78	2.88	3.66	2.55	4.40	3.54
2.18	3.43	3.84	4.60	1.76	2.85	2.27	1.38	1.24
4.22	1.67	4.51	3.50	6.69	5.92	6.42	0.95	4.57
3.10	3.26	2.95	2.45	1.95	2.06	3.10	2.03	4.98
0.86	2.61	1.92	2.78	1.01	2.64	3.14	0.86	3.68
2.71	3.06	4.33	4.63	2.47	2.92	1.38	2.81	5.71

Based upon Fisher's skewness coefficient, from which direction are the data skewed. Use the Bowley rule-of-thumb to determine the significance level.

www.ingramcontent.com/pod-product-compliance
Lightning Source LLC
Chambersburg PA
CBHW051622170526
45167CB00001B/28